I0069079

IFCoLog Journal of Logic and its Applications

Volume 3, Number 1

May 2016

Disclaimer

Statements of fact and opinion in the articles in IfCoLog Journal of Logics and their Applications are those of the respective authors and contributors and not of the IfCoLog Journal of Logics and their Applications or of College Publications. Neither College Publications nor the IfCoLog Journal of Logics and their Applications make any representation, express or implied, in respect of the accuracy of the material in this journal and cannot accept any legal responsibility or liability for any errors or omissions that may be made. The reader should make his/her own evaluation as to the appropriateness or otherwise of any experimental technique described.

© Individual authors and College Publications 2016
   All rights reserved.

ISBN 978-1-84890-214-5
ISSN (E) 2055-3714
ISSN (P) 2055 3706

College Publications
Scientific Director: Dov Gabbay
Managing Director: Jane Spurr

http://www.collegepublications.co.uk

Printed by Lightning Source, Milton Keynes, UK

All rights reserved. No part of this publication may be reproduced, stored in a retrieval system or transmitted in any form, or by any means, electronic, mechanical, photocopying, recording or otherwise without prior permission, in writing, from the publisher.

EDITORIAL BOARD

Editors-in-Chief
Dov M. Gabbay and Jörg Siekmann

Marcello D'Agostino
Natasha Alechina
Sandra Alves
Arnon Avron
Jan Broersen
Martin Caminada
Balder ten Cate
Agata Ciabttoni
Robin Cooper
Luis Farinas del Cerro
Esther David
Didier Dubois
PM Dung
Amy Felty
David Fernandez Duque
Jan van Eijck

Melvin Fitting
Michael Gabbay
Murdoch Gabbay
Thomas F. Gordon
Wesley H. Holliday
Sara Kalvala
Shalom Lappin
Beishui Liao
David Makinson
George Metcalfe
Claudia Nalon
Valeria de Paiva
Jeff Paris
David Pearce
Brigitte Pientka
Elaine Pimentel

Henri Prade
David Pym
Ruy de Queiroz
Ram Ramanujam
Chrtian Retoré
Ulrike Sattler
Jörg Siekmann
Jane Spurr
Kaile Su
Leon van der Torre
Yde Venema
Rineke Verbrugge
Heinrich Wansing
Jef Wijsen
John Woods
Michael Wooldridge

SCOPE AND SUBMISSIONS

This journal considers submission in all areas of pure and applied logic, including:

pure logical systems
proof theory
constructive logic
categorical logic
modal and temporal logic
model theory
recursion theory
type theory
nominal theory
nonclassical logics
nonmonotonic logic
numerical and uncertainty reasoning
logic and AI
foundations of logic programming
belief revision
systems of knowledge and belief
logics and semantics of programming
specification and verification
agent theory
databases

dynamic logic
quantum logic
algebraic logic
logic and cognition
probabilistic logic
logic and networks
neuro-logical systems
complexity
argumentation theory
logic and computation
logic and language
logic engineering
knowledge-based systems
automated reasoning
knowledge representation
logic in hardware and VLSI
natural language
concurrent computation
planning

This journal will also consider papers on the application of logic in other subject areas: philosophy, cognitive science, physics etc. provided they have some formal content.

Submissions should be sent to Jane Spurr (jane.spurr@kcl.ac.uk) as a pdf file, preferably compiled in LaTeX using the IFCoLog class file.

CONTENTS

Editorial Preface

Lorenzo Magnani
University of Pavia, Italy
lmagnani@unipv.it

This special issue of the *IfColog Journal of Logics and their Applications* "Frontiers of Abduction" is based on a selection of papers concerning abduction that are situated at the crossroad of logic, epistemology, and cognitive science.

The status of abduction – that Hintikka considered the fundamental problem of contemporary epistemology [7] – is very controversial. When dealing with abductive reasoning, misinterpretations and equivocations are common. What are the differences between abduction and induction? What are the differences between abduction and the well-known hypothetico-deductive method? What did Peirce mean when he considered abduction a kind of inference? Does abduction involve only the generation of hypotheses or their evaluation too? Are the criteria for the best explanation in abductive reasoning epistemic, or pragmatic, or both? How many kinds of abduction are there? The papers presented in this issue aim at increasing knowledge – logical, cognitive, epistemological – about creative and expert inferences.

Analyzing various reasoning and modeling practices provides a way of specifying the nature of some abductive reasoning processes. Following [12; 14] we have to distinguish between *sentential*, *model-based*, and *manipulative* abduction. In the past decades many attempts have been made to model abduction by developing some formal tools in order to illustrate its computational properties and the relationships with the different forms of deductive reasoning (see, for example, [4]). Some of these formal models of abductive reasoning are based on the theory of the *epistemic state* of an agent [3], where the epistemic state of an individual is modeled as a consistent set of beliefs that can change by expansion and contraction (*belief revision framework*). These formal tools, often related to the tradition of nonmonotonic logic, logic programming, semantic tableaux, dynamic and adaptive logic, etc., are usually devoted to illustrating sentential abduction and present some limitations [13]. It may be said that the traditional logical accounts of abduction certainly illustrate much of what is important in abductive reasoning, especially the objective of selecting a set of hypotheses (diagnoses, causes) that are able to dispense good (preferred) explanations of data (observations), but fail in accounting for many cases of explanations occurring in science or in everyday reasoning. The logical papers presented in this special issue aim at overcoming these limitations. Moreover, if we want to provide a suitable framework for analyzing the most interesting cases of conceptual changes in science we cannot limit ourselves to the *sentential* view of abduction but we have to consider a broader *inferential* one, encompassing both sentential and *model-based* sides of creative abduction.[1]

Many kinds of abductions involving analogies, diagrams, thought experimenting, visual imagery, etc. (for example in scientific discovery processes), can be called *model-based*. I believe that research in cognitive science, especially cognitive psychology, artificial intelligence, and computational philosophy, have established that heuristic procedures are abductive and reasoned. Among the various kinds of model-based and abductive reasoning, analogy has received particular attention from the point of view of computational models designed to simulate aspects of human analogical thinking: for example, Thagard et al. have developed ARCS (Analog Retrieval by Constraint Satisfaction; [24]) and ACME (Analogical Constraint Mapping Engine; [8]), computational programs that are built on the basis of a multiconstraint theory. Thagard and Croft have also shown that analogical reasoning is present in some forms of model-based questioning relevant to technological innovation [23]. In the last three decades many authors have demonstrated how the practices of analogical modeling, visual modeling, and thought experimenting have played generative roles in concept formation in science. They also developed several accounts of how these model-based reasoning tools function in conceptual innovation and change. For example, Craig, Nersessian, and Catrambone [5] argued that some common-sense and epistemological solutions to target problems can be generated basing on perceptual simulations of analogous but superficially dissimilar source problems.

The concept of *manipulative abduction* [17] is devoted to capturing the role of action and of external representations in many interesting situations: action provides otherwise unavailable information that enables the agent to solve problems by starting and performing a suitable abductive process of generation or selection of hypotheses [15], which occurs in a highly structured cognitive and semiotic framework [16].

Finally, the important GW and AKM schemas of abduction are clearly illustrated in [6]: indeed the classical schematic representation of abduction is expressed by what [6] call AKM-schema, which is contrasted to their own (GW-schema): *A* stands for Aliseda ([1; 2]), *K* for Kowalski ([10]), Kuipers ([11]), and Kakas *et al.* ([9]), and *M* stands for Magnani ([14]) and Meheus ([22]). On this issue cf. [18; 19], in which the recent so-called EC-model of abduction is illustrated as well.

---

[1]In [20] some model-based aspects concerning the abductive role of diagrams in mathematical thinking is addressed. An integration of the computational account of abduction with some ideas developed inside the so–called dynamical approach is illustrated in [21].

The contributions to this special issue are written by interdisciplinary researchers in logic, epistemology, and cognitive science. They aim at increasing knowledge about abductive reasoning by illustrating some of the most recent results and achievements. Three papers are centered on fundamental logical aspects: the status of logical models of abduction in the so called meta-abduction (Katsumi Inoue), the interplay between logic programming and abduction, and the application of abductive logic to morality (Luís Moniz Pereira and Ari Saptawijaya), Peirce's interrogative construal of abductive logic in a "dynamic perspective" (Minghui Ma and Ahti-Veikko Pietarinen). The remaining three papers illustrate some recent issues concerning the status of abduction with respect to logico-cognitive problems: from the methodological and computational aspects of abductive creativity in natural sciences (Rivadulla), to the role of recommendations as imperative propositions in the active operation of abductive reasoning (West), to the very recent research on the intertwining between abduction and ignorance (Bertolotti, Arfini and Magnani).

Several papers concerning abduction, practical reasoning, and scientific discovery deriving from the presentations given at the Conferences MBR (Model-Based Reasoning), promoted since 1998 by the editor of the present special issue, are contained in the following books: *Model-Based Reasoning in Scientific Discovery*, edited by L. Magnani, N.J. Nersessian, and P. Thagard (Kluwer Academic/Plenum Publishers, New York, 1999; Chinese edition, China Science and Technology Press, Beijing, 2000), which was based on the papers presented at the first "model-based reasoning" international conference, held at the University of Pavia, Pavia, Italy in December 1998. Other two volumes were based on the papers presented at the second "model-based reasoning" international conference, held at the same place in May 2001: *Model-Based Reasoning. Scientific Discovery, Technological Innovation, Values*, edited by L. Magnani and N.J. Nersessian (Kluwer Academic/Plenum Publishers, New York, 2002) and *Logical and Computational Aspects of Model-Based Reasoning*, edited by L. Magnani, N.J. Nersessian, and C. Pizzi (Kluwer Academic, Dordrecht, 2002). Another volume, *Model-Based Reasoning in Science and Engineering*, edited by L. Magnani (College Publications, London, 2006), was based on the papers presented at the third "model-based reasoning" international conference, held at the same place in December 2004. The volume *Model-Based Reasoning in Science and Medicine*, edited by L. Magnani and L. Ping (Springer, Heidelberg/Berlin 2006), was based on the papers presented at the fourth "model-based reasoning" conference, held at Sun Yat-sen University, Guangzhou, P. R. China. The volume *Model-Based Reasoning in Science and Technology. Abduction, Logic, and Computational Discovery*, edited by L. Magnani, W. Carnielli and C. Pizzi (Springer, Heidelberg/Berlin 2010), was based on the papers presented at the fifth "model-based reasoning" conference, held at the University of Campinas, Campinas, Brazil, in

December 2009. Finally, the volume *Model-Based Reasoning in Science and Technology. Theoretical and Cognitive Issues*, edited by L. Magnani, (Springer, Heidelberg/Berlin 2013), was based on the papers presented at the sixth "model-based reasoning" conference, held at Fondazione Mediaterraneo, Sestri Levante, Italy, June 2015.

Some interesting papers related to problem of abduction can be found in previous Special Issues of Journals: in *Philosophica*: Abduction and Scientific Discovery, 61(1), 1998, and Analogy and Mental Modeling in Scientific Discovery, 61(2) 1998; in *Foundations of Science*: Model-Based Reasoning in Science: Learning and Discovery, 5(2) 2000, all edited by L. Magnani, N.J. Nersessian, and P. Thagard; in *Foundations of Science*: Abductive Reasoning in Science, 9, 2004, and Model-Based Reasoning: Visual, Analogical, Simulative, 10, 2005; in *Mind and Society*: Scientific Discovery: Model-Based Reasoning, 5(3), 2002, and Commonsense and Scientific Reasoning, 4(2), 2001, all edited by L. Magnani and N.J. Nersessian; in the Special Issue of the *Logic Journal of the IGPL*: Abduction, Practical Reasoning, and Creative Inferences in Science, 14(1) (2006); in the two Special Issues of *Foundations of Science*: Tracking Irrational Sets: Science, Technology, Ethics, and Model-Based Reasoning in Science and Engineering, 13(1) and 13(2) (2008), all edited by L. Magnani. Other technical logical papers presented at MBR09_BRAZIL have been published in a special issue of the *Logic Journal of the IGPL*: Formal Representations in Model-Based Reasoning and Abduction, 20(2) (2012), edited by L. Magnani, W. Carnielli, and C. Pizzi. Finally, technical logical papers presented at MBR12_ITALY have been published in a special issue of the *Logic Journal of the IGPL*: Formal Representations in Model-Based Reasoning and Abduction, 21(6) (2013), edited by L. Magnani.

Lorenzo Magnani
University of Pavia

Pavia, Italy, January 2016

References

[1] A. Aliseda. Seeking Explanations: Abduction in Logic, Philosophy of Science and Artificial Intelligence. PhD thesis, Amsterdam: Institute for Logic, Language and Computation, 1997.

[2] A. Aliseda-Llera. *Abductive Reasoning: Logical Investigations into Discovery and Explanation*. Springer, Berlin, 2006.

[3] C. Boutilier and V. Becher. Abduction as belief revision. *Artificial Intelligence*, 77:43–94, 1995.

[4] T. Bylander, D. Allemang, M. C. Tanner, and J. R. Josephson. The computational complexity of abduction. *Artificial Intelligence*, 49:25–60, 1991.

[5] D.L. Craig, N.J. Nersessian, and R. Catrambone. Perceptual simulation in analogical problem solving. In L. Magnani and N.J. Nersessian, editors, *Model-Based Reasoning: Science, Technology, Values*, pages 167– 189, New York, 2002. Kluwer Academic/-Plenum Publishers.

[6] D.M. Gabbay and J. Woods. *The Reach of Abduction*. North-Holland, Amsterdam, 2005. Volume 2 of *A Practical Logic of Cognitive Systems*.

[7] J. Hintikka. What is abduction? the fundamental problem of contemporary epistemology. *Transactions of the Charles S. Peirce Society*, 34:503–533, 1998.

[8] K. J. Holyoak and P. Thagard. Analogical mapping by constraint satisfaction. *Cognitive Science*, 13:295–355, 1989.

[9] A Kakas, R. A. Kowalski, and F. Toni. Abductive logic programming. *Journal of Logic and Computation*, 2(6):719–770, 1993.

[10] R. A. Kowalski. *Logic for Problem Solving*. Elsevier, New York, 1979.

[11] T. A. F. Kuipers. Abduction aiming at empirical progress of even truth approximation leading to a challenge for computational modelling. *Foundations of Science*, 4:307–323, 1999.

[12] L. Magnani. Abductive reasoning: philosophical and educational perspectives in medicine. In D.A. Evans and V.L. Patel, editors, *Advanced Models of Cognition for Medical Training and Practice* , pages 21–41, Berlin, 1992. Springer.

[13] L. Magnani. Limitations of recent formal models of abductive reasoning. In A. Kakas and F. Toni, editors, *Abductive Reasoning. KKR-2, 17th International Joint Conference on Artificial Intelligence (IJCAI 2001)*, pages 34–40, Berlin, 2001. Springer.

[14] L. Magnani. *Abduction, Reason, and Science. Processes of Discovery and Explanation*. Kluwer Academic/Plenum Publishers, New York, 2001.

[15] L. Magnani. Epistemic mediators and model-based discovery in science. In L. Magnani and N. J. Nersessian, editors, *Model-Based Reasoning: Science, Technology, Values*, pages 305–329, New York, 2002. Kluwer Academic/Plenum Publishers.

[16] L. Magnani. Abduction and cognition in human and logical agents. In S. Artemov, H. Barringer, A. Garcez, L. Lamb, and J. Woods, editors, *We Will Show Them: Essays in Honour of Dov Gabbay*, pages 225–258, London, 2005. King's College Publications.

[17] L. Magnani. *Abductive Cognition. The Epistemological and Eco-Cognitive Dimensions of Hypothetical Reasoning*. Springer, Heidelberg/Berlin, 2009.

[18] L. Magnani. Is abduction ignorance-preserving? Conventions, models, and fictions in science. *Logic Journal of the IGPL*, 21(6):882–914, 2013.

[19] L. Magnani. The eco-cognitive model of abduction. Ἀπαγωγή now: Naturalizing the logic of abduction. *Journal of Applied Logic*, 13:285–315, 2015.

[20] L. Magnani and R. Dossena. Morphodynamical abduction: causation by attractors dynamics of explanatory hypotheses in science. *Foundations of Science*, 10:107–132, 2005.

[21] L. Magnani and M. Piazza. Perceiving the infinite and the infinitesimal world: unveiling and optical diagrams and the construction of mathematical concepts. *Foundations of Science*, 10:7–23, 2005.

[22] J. Meheus, L. Verhoeven, M. Van Dyck, and D. Provijn. Ampliative adaptive logics and the foundation of logic-based approaches to abduction. In L. Magnani, N. J. Nersessian, and C. Pizzi, editors, *Logical and Computational Aspects of Model-Based Reasoning*, pages 39–71. Kluwer Academic Publishers, Dordrecht, 2002.

[23] P. Thagard and D. Croft. Scientific discovery and technological innovation: ulcers, dinosaur extinction, and the programming language java. In N.J. Nersessian L. Magnani and P. Thagard, editors, *Model-Based Reasoning in Scientific Discovery*, pages 125–137, New York, 1999. Kluwer Academic/Plenum Publishers.

[24] P. Thagard, K. J. Holyoak, P. Nelson, and G. Gochfeld. Analog retrieval by constraint satisfaction. *Artificial Intelligence*, 46:259–310, 1990.

META-LEVEL ABDUCTION

KATSUMI INOUE
National Institute of Informatics, Japan
Tokyo Institute of Technology, Japan
inoue@nii.ac.jp

Abstract

Meta-level abduction (MLA) has been proposed as a method to abduce missing laws in completing proofs and explaining observations at the meta-level. Based on a simple logic of causality, Inoue *et al.* (2010) firstly proposed meta-level abduction to discover physically unobserved causality in terms of *hidden rules* to explain given *empirical rules* with respect to skills for music playing. Meta-level abduction has also been applied to completion of biological networks containing both positive and negative causal effects (Inoue *et al.*, 2013). In this paper, we define a general framework for meta-level abduction together with a logical system for it, and analyze its potential power in various patterns of abductive reasoning. We will see that meta-level abduction can realize second-order existential abduction by Schurz (2008). Moreover, meta-level abduction can be coordinated with selective, creative and other types of abductions.

1 Abduction and Meta-Reasoning

1.1 Abductive Reasoning

Abduction is one of the three fundamental modes of reasoning characterized by Peirce [58], the others being deduction and induction. Abduction amounts to concluding the minor premise from the major premise and the conclusion, and its original inferential style is given as follows.

> The (surprising) fact, ψ, is observed;
> But if φ were true, ψ would be a matter of course;
> Hence, there is reason to suspect that φ is true.

This corresponds to the rule of the form, called the *fallacy of affirming the consequent*:

$$\frac{\psi \quad \varphi \to \psi}{\varphi} . \tag{1}$$

φ is called an *explanans* for an *explanandum* ψ.

In the field of Artificial Intelligence (AI), logic of abduction has often been studied from the point of view of automated reasoning [56, 29] and its applications to diagnosis and recognition [35]. There, it is often assumed in (1) that both φ and ψ are propositional or first-order formulas and the arrow $\rightarrow$ is interpreted as an implication connective in such a logic. The major premise $(\varphi \rightarrow \psi)$ is either given in a prior (or *background*) *knowledge* Σ, i.e., $(\varphi \rightarrow \psi) \in \Sigma$, or is inferred from it, i.e., $\Sigma \models (\varphi \rightarrow \psi)$. In either case, the inference rule (1) can be expressed in the *meta-theoretical relation*:

$$\Sigma \wedge \varphi \models \psi. \tag{2}$$

Under these assumptions,[1] abductive proof procedures have been designed in AI, often based on deduction with the *resolution principle* [67], e.g., [63, 14, 62, 11, 28, 74, 36, 65]. From the theorem-proving viewpoint, abduction augments sufficient conditions that are missing in the premises, i.e., background knowledge Σ, to enable a derivation, i.e., proof, of the given observation ψ. This inference fills the gap in a proof of the observation from the premises. An inferred sufficient condition φ is called a *hypothesis* or an *explanation*. Note that a hypothesis can be any formula, e.g., a conjunction (set) of atoms, literals or rules. Moreover, a proof system or a consequence relation $\models$ can be based on any classical or non-classical logic.

Abduction plays essential roles in *knowledge discovery* in science and technology [45]. Then, abduction has been used since 2000s in the field of *inductive logic programming* (ILP), which originally focused on induction of logic programs [51]. The use of background knowledge in scientific applications has directed an attention of ILP to *theory completion* [49] rather than classical learning tasks such as concept learning and classification. There, abduction is mainly used to complete proofs of observations from *incomplete* background knowledge, while induction refers to generalization of the abduced cases.

In any case, abductive inference by way of (2) is *meta-theoretical*, that is, logic of abduction is given at the meta-level.

1.2 Meta-Reasoning

On the other hand, *meta-reasoning* has been intensively investigated in AI or logic programming from a distinct viewpoint from abduction. Meta-reasoning has been used for the meta-level control of reasoning, thereby enabling us the introspective monitoring of the reasoning process at the object level, e.g., [59, 8, 68]. Meta-reasoning has also been applied to *meta-programming* in logic programming, e.g., [7, 24, 38, 23, 13]. Bowen and Kowalski's

---

[1] Usually, the consistency relation that $\Sigma \wedge \varphi$ is consistent is also required. Additionally, φ is often to be as minimal as possible by virtue of Occam's Razor.

meta-interpreter [7] and its "Vanilla" variant [24] have been most widely used to implement meta-logic in logic programming, which are given as:

$$solve(true).$$
$$solve(A \wedge B) \quad \leftarrow \quad solve(A) \wedge solve(B).$$
$$solve(\neg A) \quad \leftarrow \quad \textbf{not } solve(A).$$
$$solve(A) \quad \leftarrow \quad rule(A \leftarrow B) \wedge solve(B).$$

Here, variables are capitalized in logic programming, e.g., A, B and C,[2] and **not** represents *negation as failure* [10]. The predicates *solve* and *rule* are *meta-predicates*, and all constructs with them are atoms. An atom of the form $solve(A)$ reflects that A is proved from the program, and $rule(A \leftarrow B)$ denotes that $(A \leftarrow B)$ is a rule in the program.[3] Note that a *fact* of the form A can be expressed as $rule(A \leftarrow true)$ in a meta-program. In general, any inference rule can be expressed as such a meta-rule, e.g.,

$$solve(A \leftarrow B) \quad \leftarrow \quad solve(A \leftarrow C) \wedge solve(C \leftarrow B).$$
$$solve(\neg A) \quad \leftarrow \quad solve(B \leftarrow A) \wedge solve(\neg B).$$

Note that those meta-level axioms have been used for *deduction* in meta-programming or *deduction control* [59].

1.3 Meta-Level Abduction

Now, a question arises. What happens if abduction is applied for such meta-level axioms? For example, suppose that the meta-rule representing Modus Ponens (MP)

$$solve(A) \leftarrow rule(A \leftarrow B) \wedge solve(B)$$

as well as the meta-level expression of a minor premise q

$$solve(q),$$

---

[2]Those free variables are assumed to be universally quantified at the front of each rule. Conventionally, the connective $\leftarrow$ is used in logic programming instead of $\rightarrow$ as material implication, that is, $(A \leftarrow B)$ and $(B \rightarrow A)$ is logically equivalent.

[3]The connective symbol $\leftarrow$ in the atom $rule(A \leftarrow B)$ is used here for notational convenience. Instead, an atom of the form $rule(A \leftarrow B)$ should be expressed as $rule(A, B)$ in logic programming. The meta-predicates *solve* and *rule* are originally represented as *demo* and *clause*, respectively. The meta-predicate *demo* takes another argument T to represent a theory in the meta-interpreter in [7] and $demo(T, A)$ represents that A is provable in T, while the Vanilla meta-interpreter [24] omits the theory argument. The meta-predicate *clause* has two arguments as $clause(A, B)$, which represents the rule $(A \leftarrow B)$.

are contained in a background theory Σ. Given the meta-level expression of a consequent p as an observation

$$solve(p),$$

we can abduce by (2)

$$rule(p \leftarrow q).$$

In this example, $(p \leftarrow q)$ is a rule, hence *law abduction* [45] is realized, yet this is an ordinary abduction in AI, called *fact abduction* in philosophy [71, 27], since it only abduces the atom with the meta-predicate $rule$. Inoue *et al.* have called this inference *meta-level abduction* (MLA) [32, 31], since abduction is performed at the meta-level rather than at the object level. It is not called "meta-abduction" because meta-abduction would mean abduction about abduction, cf., [4]. Meta-level abduction is abduction at the meta-level and thus is abduction to augment what the meta-theory is intended to infer, but could be also applied for meta-abduction if the meta-level axioms are given for abduction.

It is notable that most previous abductive methods in the AI literature deal with fact abduction. Yet, it is possible to implement law abduction by providing a predetermined set of candidate rules (called *abducible* rules). Previous abductive methods of abduction thus select proper combinations of abducible facts or abducible rules. In this sense, abduction in AI is generally *selective* [45, 46]. In contrast, meta-level abduction does not need any such abducible rules in advance.

Meta-level abduction has been applied to discover physical skills in terms of *hidden rules* to explain given *empirical rules* in the domain of cello playing in [32]. By representing rule structures of the problem in a form of *causal networks*, meta-level abduction infers missing links and unknown nodes from incomplete networks to complete paths from stimuli to effects. Then, applicability of meta-level abduction to deal with networks expressing both positive and negative causal effects has been examined in [31]. Such negative causal effects are known in biology as *inhibitory effects*, which are essential in gene regulatory, signaling and metabolic networks. Such biological applications of meta-level abduction to networks with inhibition have been reported in treatment of hypertension [41], completion of p53 signaling networks [31] and completion of glucose repression networks [77]. Meta-level abduction has also been extended to realize *analogical abduction* by adding meta-axioms for analogy, which has been applied to cello exercises to obtain analogical explanations for skill acquisition [20].

1.4 Related Research

Meta-reasoning has been intensively applied to logic programming [38, 23, 13], and it is possible to implement abductive procedures using meta-programming to perform abduction [74, 17, 36, 2]. Such abductive meta-interpreters should be distinguished from abduction at the meta-level, and law abduction had not been considered in the literature of meta-reasoning before MLA was proposed in [32]. For example, Kowalski discussed the importance of both abduction and meta-reasoning in [38], but did not mention their combination there. Christiansen [9] computes both abduction and induction in a unified system of meta-programming, and uses separate forms of reasoning in their actual computation followed by the system. In [9], the *demo* predicate is used to generate parts of programs that are necessary to derive the goal based on techniques of constraint logic programming.

Muggleton *et al.* [52, 53] propose *meta-interpretive learning* (MIL) for law abduction. Similar to MLA, MIL uses an abductive meta-interpreter for generating rules from the pre-stored rule schemas. However, the use of meta-interpreters in MIL is different from that in MLA; MLA applies abduction for meta-level axioms, but MIL uses abduction for object-level theories with higher-order abducible rules. *Predicate invention* [50] is an important method in ILP to introduce new objects and relations, and both MLA and MIL have implemented predicate invention by abduction but in different ways. MLA in [32] introduces new objects or propositions as new terms of meta-predicates, but MIL in [52] introduces new symbols which represent relations.

Some attempts in AI and ILP have contributed to abduction and induction in *causal theories*, which are formally represented as sets of individual rules between causes and their effects. The *event calculus* [39] is a meta-theory for reasoning about time and action in the framework of logic programming. *Abductive event calculus* [15] is an abductive extension of the event calculus, and can be regarded as a kind of meta-level abduction. Abductive event calculus has been extended for applications to planning, e.g., [72], but has never been used for abducing causal theories.

Bharathan and Josephson [4] use a specific structure of abduction that is tailored for belief revision, then meta-abductive reasoning is performed over the records of an agent's reasoning trace to make requisite changes to beliefs. Such a meta-abduction reasons about reasoning states and steps as a controller of abduction, whereas our MLA performs abduction about observations using a meta-theory as a background knowledge. Pereira and Pinto [60] investigate side-effects of interest called *inspection points* in choosing abductive solutions. Inspection points can be embedded as meta-predicates on top of existing abductive systems. Similar to [4], this is a work on controlling abduction at the meta-level.

Gabbay and Woods [21] show how to make an abductive logic based on some *base logic* and the labeled deduction system. In this sense, the logic of abduction is discussed at the meta-level, thereby formalizing abduction in a flexible manner upon any base logic. MLA can also be based on any base logic and can be applied to meta-level axioms of such a logic.

There are several attempts to propose meta-level inference rules (or *structural rules*) that represent rationality postulates for abductive/explanatory relations, thereby enabling a logical characterization of abduction at the meta-level [18, 48, 42, 3, 6]. Compared with these works, MLA does not offer a logic for abduction itself, since MLA just applies abduction to a meta-theory in any base logic using any abductive computation for that logic. It would thus be possible to write such structural rules for abduction as meta-level axioms and then to perform MLA upon them. Such a possibility will be further discussed later in Section 3.2.

In the rest of this paper, we analyze MLA by giving a general framework that subsumes specific MLA systems in [32, 31]. Based on this new MLA framework, we reveal the potential inference power of MLA and give a general perspective of MLA towards creative abduction.

2   Patterns of Abduction

Our goal is to see the potential power of meta-level abduction from the point of view of forms of explanans and explanundums. To this end, this section reviews the patterns of abduction investigated by Hirata [26], Magnani [45, 46], Schurz [71] and Hoffmann [27], which are then utilized in a mixed way in later sections.

Hirata [26] classifies abduction into 5 types in the context of logic programming: *rule-selecting*, *rule-finding*, *rule-generating*, *theory-selecting*, and *theory-generating* abductions. This classification is based on works on abduction in AI and abductive and inductive logic programming, and focuses mainly on how hypotheses are obtained with or without background theories.

Magnani [45, 46] gives a wider perspective on different epistemological types of abduction, and in particular shows two distinct notions of abduction, i.e., *selective* and *creative* abductions. The task in selective abduction is to abduce hypotheses by selecting them from an encyclopedia of pre-stored entities in the language of hypotheses, while a discovery of new hypotheses is in order in creative abduction.

Schurz [71] then further classified patterns of abduction broadly into 4 kinds: *factual*, *law*, *theoretical-model*, and *second-order existential* abductions, which are further divided into 8 patterns in Table 1. Both theoretical-model abduction and 2nd-order existential abduction are kinds of creative abduction discussed by Magnani. According to Schurz, theoretical-model abduction introduces a new *theoretical model* that consists of facts, laws and concepts without using any new facts and concepts, that is, the organization and combination of its components are new, while the components themselves are already known. Classification among 2nd-order existential abduction is based on the situations in which abduction is performed, and all patterns in it generate new laws with new concepts.[4] Schurz

---

[4]In [71], "Hypothetical cause abduction" is further divided into "Speculative abduction", "Strict common cause abduction", "Statistical factor analysis", and "Abduction to reality". Schurz also specified in his classification how each abduction is driven, which is omitted in Table 1.

Kind of abduction	Explanandum	Explanans
Factual abduction	Single facts	New facts
— Observable-fact abduction	Single facts	Factual reasons
— 1st-order existential abduction	Single facts	Facts with new unknown individuals
— Unobservable-fact abduction	Single facts	Unobservable facts
Law abduction	Empirical laws	New laws
Theoretical-model abduction	General empirical phenomena	New theoretical models
2nd-order existential abduction	General empirical phenomena	New laws with new concepts
— Micro-part abduction	General empirical phenomena	Microscopic compositions
— Analogical abduction	General empirical phenomena	New laws with analogical concepts
— Hypothetical cause abduction	General empirical phenomena	Hidden (unobservable) causes

Table 1: Patterns of abduction by Schurz (part)

Explanans	Exists in our mind	Exists in our culture / Historically new
Facts	Selective fact abduction	P-creative/H-creative fact abduction
Types (or concepts)	Selective type abduction	P-creative/H-creative type abduction
Laws	Selective law abduction	P-creative/H-creative law abduction
Theoretical models	Selective model abduction	P-creative/H-creative model abduction
Representations	Selective meta-diagrammatic abduction	P-creative/H-creative meta-diagrammatic abduction

Table 2: Hoffmann's taxonomy of kinds of abduction

considered both factual and law abductions are selective abduction, but the distinction between factual and law abductions are mainly with regard to the form of explanations.[5]

Hoffmann [27] then argued that factual and law abductions can be creative too, so this syntactic distinction should be orthogonal to that of selective and creative abductions. Hoffmann's *fact abduction* is similar to Schurz's factual abduction, but *type abduction* is distinguished from it to denote abduction of new concepts. *Law abduction* by Hoffmann can also include new concepts. Table 2 is the summary of patterns of abduction that are modified by Hoffmann.[6] Hoffmann's *meta-diagrammatic abduction* partially correspond to Schurz's 2nd-order existential abduction, but is mostly creative, since completely new theoretical models are required by changing or developing new representation systems.

2.1 Existing Abductive Systems

Before we go deep into the characterization of meta-level abduction (MLA), here we connect past works on abductive inference systems with patterns of abduction.

---

[5]Roughly speaking, Hirata's rule-selecting and rule-finding abductions in [26] correspond to Schurz's factual abduction and analogical abduction, respectively. On the other hand, rule-generating and theory-selecting abductions in [26] correspond to law abduction. Hirata's theory-generating abduction is more like (explanatory) induction in ILP, in particular is equal to Shapiro's *model inference system* [73].

[6]For creative abduction, "P" stands for "psychological", and "H" stands for "historical" [27].

As mentioned in Section 1.3, most abductive procedures which are developed in AI and logic programming deal with fact abduction only. In particular, those resolution-based abductive procedures, e.g., [63, 14, 62, 28, 74, 36], are of this kind. First-order existential abduction by Schurz has been realized in such procedures as in [63, 14, 28, 65] by computing hypotheses with existentially quantified variables. Other types of abductive computation are also mostly considered for fact abduction or selective abduction [35]. Console *et al.* [11] have shown an object-level characterization of meta-level expression (2) in terms of deduction and completion. This work uses deduction for fact abduction.

Some works on law abduction had existed in AI before MLA was introduced in [32]. In most works, a set of candidate rule patterns called *abducible rules* must be pre-stored in advance, hence they are all instances of selective abduction. The framework of abductive systems with such abducible rules was firstly considered by Poole [61], which associates a unique name with each ground instance of an abducible rule schema, and those rule names are abduced by fact abduction. This is feasible when we know exact patterns of rules as strong biases, and offers a convenient method to reduce selective law abduction to fact abduction. However, it is impossible to prepare all patterns of rules in advance in order to abduce missing rules.

Inductive Logic Programming (ILP) [51] is a subfield of machine learning in AI which uses logic programming as a uniform representation for (training) examples/observations, background knowledge and hypotheses. As its name indicates, ILP aims at development of theory and practice of induction, but it has been recognized that *explanatory induction* can be viewed as abduction [19], in particular as law abduction. An early work by Shapiro called *Model Inference System* [73] builds a theory from scratch within a given language, and can thus be viewed as theoretical-model abduction. *Predicate invention* [50] had been intensively investigated in the initial stage of ILP research and has recently been revisited [51, 53], since it should play an important role in discovery. Predicate invention itself corresponds to type abduction by Hoffmann, and ILP with predicate invention partially realizes creative abduction, since it introduces new concepts and new laws.

Fact abduction has been used to implement law abduction in several ILP systems. Including [69] and other works in [19], most such previous systems use abduction to compute inductive hypotheses cleverly, and such integration is useful in theory refinement. CF-induction [30] can induce explanatory rules as hypotheses, hence realizes law abduction in first-order clausal theories. CF-induction can directly induce first-order full clausal theories at the object level based on *inverse entailment*, viz., $\Sigma \wedge \neg \psi \models \neg \varphi$ for (2), but predicate invention should be realized by *inverse resolution* [50] and the search space for hypothesis enumeration is huge in general. TAL [12] translates an ILP problem into an equivalent abductive problem, then an abductive procedure is used to abduce a hypothesis for an observation from the background theory and a pre-stored inductive bias called a *top theory*. TAL uses the naming method to abduce rules as in [61]. *Meta-interpretive learning* (MIL)

[52, 53] is a further refined ILP method to generate rules from the pre-stored rule schemas, and hence realizes law abduction in a selective way. Using MIL, predicate invention as well as learning of recursive programs can be efficiently implemented by way of abduction with respect to a meta-interpreter of logic programming.

MLA was introduced in [32] as a method to discover unknown relations from incomplete knowledge bases. The main objective to use MLA in [32] is to provide knowledge representation and a reasoning method for law abduction. MLA in [32] and its nonmonotonic extension [31] have been implemented in SOLAR [54, 55], an automated deduction system for *consequence finding* [40, 28, 47] in first-order logic. Meta-level abduction by SOLAR is powerful enough to infer missing rules, missing facts, and unknown causes involving predicate invention [50] in the form of existentially quantified hypotheses.

2.2   Lifting to the Meta-Level

Based on the summary of abductive computation in the last subsection, we should choose an appropriate procedure for a pattern of abduction. Most methods are for selective abduction, and abductive procedures developed in AI and logic programming are for fact abduction. Law abduction has been implemented in ILP, and predicate invention introduces new concepts. Some ILP works can be considered to realize creative abduction. In the theoretical-model abduction by Schurz [71], multiple laws and facts are newly combined to form a hypothesis to explain multiple observations. Hence, from the computational viewpoint, we do not have to distinguish theoretical-model abduction from law abduction. In Section 3, we will see that MLA can realize 2nd-order existential abduction by Schurz [71] at the object level, although MLA in [32] itself can be classified as fact abduction (in the sense of Hoffmann [27]) at the meta-level, which we refer to as *fact meta-level abduction* hereafter.

The last observation allows us to expect that meta-level abduction can lift the type of abduction to one level higher. After we verify that fact meta-level abduction can actually realize creative law abduction at the object level in Section 3, we will consider *law meta-level abduction* and its potential usage in Section 4. Then, we will conclude that MLA can be coordinated with selective, creative and other types of abductions by [45].

3   Fact Meta-Level Abduction

Apart from the specific form of MLA in [32, 31], we here give a general definition of MLA. We suppose a language $\mathcal{L}$ that includes both a meta-level language $\mathcal{L}_M$ and an object-level language $\mathcal{L}_O$ as well as their *amalgamation* [7]. The language $\mathcal{L}_M$ contains higher-order (meta-)predicates and predicate/function/constant variables, while the language $\mathcal{L}_O$ contains predicates, functions, constants and variables at the object level. It is important here that the domain of the language $\mathcal{L}$ is assumed to contain all possible individuals, but some of them are not practically stored in the domain in advance. That is, new individuals and new relations can be introduced in the language when they become necessary. We then suppose a *meta-theory* T in $\mathcal{L}$, which is composed of elements in $\mathcal{L}_M$ with constructs from $\mathcal{L}_O$ in their arguments. We also assume a consequence relation $\models$ over $\mathcal{L}$.

Definition 3.1. An *MLA framework* is defined as a triple $(T, \mathcal{A}, G)$, where T is a meta-theory in $\mathcal{L}$, $\mathcal{A}$ is a set of formulas in $\mathcal{L}$ called *abducibles* and G is a formula in $\mathcal{L}$ called an *observation* or a *goal*. An abductive task is to find an *explanation* E of G such that (i) E is a conjunction of abducibles from $\mathcal{A}$, and (ii) $T \wedge E \models G$.[7]

Definition 3.2. A theory for *fact MLA* is an MLA framework $(T, \mathcal{A}, G)$, in which $\mathcal{A}$ is a set of facts from $\mathcal{L}$.

As case studies of fact MLA, Section 3.1 will review an MLA framework introduced in [32] for causal networks, and Section 3.2 will provide a new logical characterization of causal networks and abduction on them. Section 3.3 will revisit an extension of MLA for nonmonotonic causal networks introduced in [31]. All these MLA frameworks are analyzed to see their abductive power with regard to patterns of abduction.

3.1 Simple Logic of Causality

In [32], a background theory is assumed to be represented in a network structure called a *causal network*. A causal network is a directed hyper-graph, which consists of a set of *nodes* and a set of (directed) *(hyper-)arcs* (or *links*). Each node in a causal network represents some event, fact or proposition. A *direct causal link* corresponds to a directed arc, and a *causal chain* is represented by the reachability through direct causal links between two nodes.[8] A first-order language can be associated to express causal networks as follows. A node is associated with a *proposition* or a (ground) *atom* in the object-level language. A link is associated with a (ground) *rule* in the object-level language. Namely, when there is a direct causal link from a node s to a node g, it is expressed as a rule of the form $(g \leftarrow s)$. In this case, we define that $linked(g, s)$ is true as in (3) at the meta-level, where $linked$ is a meta-predicate.[9]

$$g \longleftarrow s \qquad\qquad linked(g, s) \qquad\qquad (3)$$

---

[7]As in Footnote 1, the consistency of $T \wedge E$ as well as the minimality of E are usually required.

[8]The interpretation of a "cause" here just represents the connectivity, which may refer to any dependency such as a physical, chemical, conceptual, epidemiological, social, structural, or statistical dependency [57]. A "direct cause" here simply represents the adjacent connectivity in a causal graph. In general, treatment of causality requires tackling many important issues that have been discussed in such as [25, 76, 57], but we do not intend to argue what it should be in this paper. Instead, we here use the notion of causality as a means of connecting observations and abductive explanations.

[9]The meta-predicate $linked$ is expressed as *connected* in [32].

The statement that "g is *jointly* caused together directly by s and t", written as $(g \leftarrow s \wedge t)$ at the object level, is expressed in a disjunction of the form (4) at the meta-level (see Equation (7) for why the disjunction is used here instead of conjunction).

$$linked(g, s) \vee linked(g, t) \tag{4}$$

In the above expression, (i) each atom at the object level is represented as a *term* at the meta-level, and (ii) each rule, i.e., a causal link, at the object level is represented as a (*disjunctive*) *fact* at the meta-level. For object-level propositions g and s, $caused(g, s)$ is defined true if there is a *causal chain* from s to g, where *caused* is another meta-predicate.[10] The $caused/2$ relation is recursively defined with the $linked/2$ facts as follows.

Definition 3.3. A theory of *Simple Logic for Causality* (SLC) is represented as a meta-theory T_{SLC} that consists of a set of meta-facts of the form (3) or (4) mentioning object-level direct causal links, together with the two meta-rules:[11]

$$caused(X, Y) \leftarrow linked(X, Y). \tag{5}$$

$$caused(X, Y) \leftarrow linked(X, Z) \wedge caused(Z, Y), \tag{6}$$

A theory for *SLC-based MLA* is an MLA framework $(T_{SLC}, \mathcal{A}_{SLC}, G)$, in which $\mathcal{A}_{SLC}$ is a set of disjunctive facts with the meta-predicate $linked$ and G is a disjunctive fact with the meta-predicate $caused$.

Law abduction can be realized in an SLC-based MLA as follows. Suppose an observation (or a goal) G represented as a causal chain $caused(g, s)$, we want to know why (or how) G can be caused. G corresponds to an *empirical rule* to be explained if it is an observation or represents a *virtual goal* to be achieved if it is a goal. An abductive task is then to find *hidden rules* that establish a connection from the source s to the target g by filling the gaps in causal networks. For this, an explanation consisting of missing causes (links) and missing facts (nodes) for the causal network are abduced in an SLC-based MLA.

For example, suppose the observation $O = caused(g, s) \wedge caused(h, s)$, that is, the multiple causal chains between two goal facts g, h and the source fact s. Examples of

---

[10]The meta-level expression $caused(g, s)$ can be related to the *input-output pair* (s, g) in [43] or the *production rule* $(s \Rightarrow g)$ in [5].

[11]The notation for meta-theories here is the same as that for logic programming.

minimal explanations of O containing two intermediate nodes are as follows.

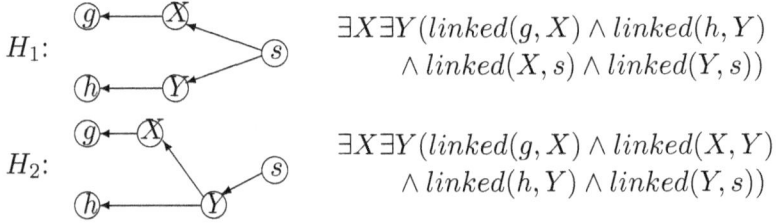

$H_1:$
$$\exists X \exists Y (linked(g, X) \wedge linked(h, Y)$$
$$\wedge \, linked(X, s) \wedge linked(Y, s))$$

$H_2:$
$$\exists X \exists Y (linked(g, X) \wedge linked(X, Y)$$
$$\wedge \, linked(h, Y) \wedge linked(Y, s))$$

H_1 and H_2 represent different connectivities, and we may want to enumerate different types of network structures that are missing in the original causal network. Moreover, these hypotheses contain existentially quantified variables, where X and Y are newly invented here. Those new terms can be regarded as either some existing nodes or new unknown nodes. Since new formulas can be produced at the object level, *predicate invention* [50] is partially realized here.[12]

Connecting the meta-level and the object-level representation, meta-literals in SLC have been translated to first-order logic formulas at the object level in [32]: An atom $linked(s, t)$ at the meta-level corresponds to a rule $(s \leftarrow t)$ at the object level. Then, a disjunction $linked(g, s) \vee linked(g, t)$ (4) corresponds to

$$linked(g, s) \vee linked(g, t) \quad \Leftrightarrow \quad (g \leftarrow s) \vee (g \leftarrow t) \quad \equiv \quad (g \leftarrow s \wedge t). \qquad (7)$$

Based on this translation, the soundness and completeness of law abduction in SLC-based MLA can be derived. Given a meta-theory T_{SLC} in SLC, let $\lambda(T_{SLC})$ be the object-level theory obtained by replacing every $linked(t_1, t_2)$ (t_1 and t_2 are terms) appearing in $T_{SLC} \setminus \{(5), (6)\}$ as (disjunctive) facts with the formula $(t_1 \leftarrow t_2)$. Note that replacement by λ in T_{SLC} does not apply to the meta-axioms (5) and (6) as these are subtracted, hence $\lambda(T_{SLC})$ does not contain any meta-predicate.

Proposition 3.0.1. [31, Theorem 1] *Suppose an SLC-based MLA* $(T_{SLC}, \mathcal{A}_{SLC}, G)$. *Let* G *be a meta-formula of the form* $G = (caused(g, s_1) \vee \cdots \vee caused(g, s_n))$. *Let* E *be an explanation of* G *constructed from* $\mathcal{A}_{SLC}$. *Then,*

$$T_{SLC} \wedge E \models G \quad iff \quad \lambda(T_{SLC}) \wedge \lambda(E) \models (g \leftarrow s_1 \wedge \cdots \wedge s_n).$$

Besides the use in law abduction, SLC-based MLA can also be applied to fact abduction [32]. Abduction of facts at the object level can be formalized as *query answering* at

---

[12]One could argue that this form of predicate invention can be regarded as a realization of *object invention* rather than *relation invention*. Since a concept can be represented as an object, type or relation, it can be regarded as a *concept invention*, yet we use predicate invention as a conventional term in ILP.

the meta-level. Given a goal of the form $caused(g, X)$, abduction of causes is computed by answer substitutions to the variable X. To perform selective fact abduction, an abducible literal a at the object level is associated with the fact $caused(a, a)$ at the meta-level. Finally, law abduction and fact abduction can be combined in the form of *conditional query answering* [33], which extracts answers in a query with additional abduced conditions. Thus, SLC-based MLA enables us to abduce both rules and facts [32].

In summary, introduction of new facts, new rules along with new concepts by SLC-based MLA indicates that MLA is able to abduce new theoretical models within SLC. In other words, SLC-based MLA realizes 2nd-order existential abduction at the object-level by performing abduction in SLC.

3.2 General Logic of Causality

Here we consider a generalization of SLC called the *general logic of causality* (GLC). In the correspondence between SLC-based MLA and object-level abduction, each direct cause of the form $linked(g, s)$ is translated to $(g \leftarrow s)$ in SLC. The material implication $\leftarrow$ has the property of transitivity, which corresponds to transitivity of causality via (5) and (6) in a causal network. Another device in SLC is that a conjunctive causal effect (4) is translated to a disjunction of the form (7), which also depends on the classical interpretation of $\leftarrow$.

Under GLC, we should not consider any particular interpretation of the meta-predicate *linked* so that its semantics is left open. This is generally desirable, since causality cannot be interpreted as material implication. Instead, the meaning of causality at the object level is now abstracted away. Once a set of *linked* atoms are given as a causal network, the meaning of the chained causality predicate *caused* should be determined by the postulates that are written in inference rules.

In GLC, we will write a conjunctive cause (4) as a direct conjunctive cause of the form $linked(g, s \wedge t)$ by allowing any conjunction in an argument of meta-facts, thereby avoiding a disjunctive fact of the form (7) used in SLC. More generally, a causal network can be given as a set of atoms of the form $linked(F, G)$, where F and G are Boolean formulas. Similarly, we can allow atoms of the form $caused(F, G)$ with Boolean formulas F and G in GLC.

To propose the postulates for causal networks in GLC, let us denote the meta-atoms $linked(g, s)$ and $caused(g, s)$ as $(s \mapsto g)$ and $(s \vDash g)$, respectively, by simply reversing the order of two arguments. Let $\succeq$ be a standard Tarski consequence relation such that $\alpha \succeq \beta$ iff α logically entails β. Then, the logic of GLC is then expressed in the following inference rules.

$$(\text{Reflection}) \quad \frac{\alpha \mapsto \psi}{\alpha \vDash \psi} \tag{8}$$

$$(\text{Weakening}) \quad \frac{\alpha \vDash \beta \quad \beta \succeq \psi}{\alpha \vDash \psi} \tag{9}$$

$$(\text{Strengthening}) \quad \frac{\alpha \succeq \beta \quad \beta \vDash \psi}{\alpha \vDash \psi} \tag{10}$$

$$(\text{And}) \quad \frac{\alpha \vDash \varphi \quad \alpha \vDash \psi}{\alpha \vDash \varphi \wedge \psi} \tag{11}$$

$$(\text{Cut}) \quad \frac{\alpha \vDash \beta \quad \alpha \wedge \beta \vDash \psi}{\alpha \vDash \psi} \tag{12}$$

The rule of Reflection (8) corresponds to the meta-axiom (5) in SLC, and is the means to connect the direct causal links $\mapsto$ with the causal relation $\vDash$, while all other causal statements are inferred by other inference rules. Note that we do not include the inference rule:

$$\frac{\alpha \succeq \psi}{\alpha \mapsto \psi}$$

and in particular $(\alpha \mapsto \alpha)$ is not satisfied unless it is explicitly given as a self-loop in the input network. This is because the consequence relation is not generally considered as a causal relation and is in fact unnecessary for our purpose. However, a consequence relation is useful for weakening and strengthening a causal statement. The rule of (Right) Weakening (9) often appears in other consequence relations, and the rule of (Left) Strengthening (10) is also called Monotonicity [37]. The Cut rule (12) originated in Gentzen systems and is also an abbreviation of Cumulative Transitivity [37]. Note that Cut and Strengthening imply the transitivity:[13]

$$(\text{Transitivity}) \quad \frac{\alpha \vDash \beta \quad \beta \vDash \psi}{\alpha \vDash \psi} \tag{13}$$

If necessary, we can introduce more inference rules. For example, if a nondeterministic effect is given in the form of $linked(g \vee h, s)$[14] or a disjunctive cause is put inside a link

---

[13]There are pros and cons for the claim that causation is transitive. On the negative side, see [25] for some examples. Moreover, the monotonicity by Strengthening apparently fails to hold in nonmonotonic logics, which implies that GLC is not always applicable. We do not argue the appropriateness of these properties here, but can merely think that the causality notion in SLC/GLC is nothing but the reachability in a causal network defined via the relation $linked$. Note that the problem of nonmonotonicity is partly handled in Section 3.3.

[14]The meta-atom $linked(g \vee h, s)$ in GLC can be expressed as the disjunction $linked(g, s) \vee linked(h, s)$ in SLC [32].

like $linked(g, s \lor t)$, we need an inference rule to handle disjunctions:

$$(\text{Or}) \quad \frac{\alpha \vDash \psi \quad \beta \vDash \psi}{\alpha \lor \beta \vDash \psi} \tag{14}$$

Using this inference system, we can deduce all rational causal statements. Here, causality in GLC is viewed as a deductive consequence relation, rather than an abductive consequence relation in such as [18, 42],[15] and its logical concept is somewhat similar to logics in [43, 6],[16] although these works are related to each other. The next proposition shows that GLC can derive all causal statements that are derived by SLC. Since GLC allows more general statements like $caused(p \land q, r \lor s)$, GLC is stronger (and is thus more general) than SLC.

Proposition 3.0.2. *Suppose a theory T for SLC, and let T' be a theory for GLC obtained from $T \setminus \{(5), (6)\}$ by replacing every disjunctive fact of the form $(linked(p, q_1) \lor \cdots \lor linked(p, q_n))$ with a meta-fact of the form $(q_1 \land \cdots \land q_n \mapsto p)$. If $G = (caused(g, s_1) \lor \cdots \lor caused(g, s_n))$ is derived from T in SLC, then $G' = (s_1 \land \cdots \land s_n \vDash g)$ is derived from T' in GLC.*

Proof. In GLC, Reflection (8) corresponds to the meta-axiom (5) in SLC. Moreover, the meta-axiom (6) in SLC can be obtained by Transitivity (13) (which is a consequence of Strengthening (10) and Cut (12)) and Reflection. Hence, the two meta-axioms in SLC also hold in GLC, and all inferences with these axioms in SLC can be obtained in GLC.

Next we consider inference involving disjunctive facts in SLC. This involves the case inference for each disjunct in a disjunction, and we can show that the corresponding inference is obtained in GLC. Although a complete proof needs a structural induction, we here show an example in the case of 2 disjuncts. Suppose a conjunctive direct link represented by $linked(q, p_1) \lor linked(q, p_2)$ in SLC. From $caused(p_1, s_1)$ and $caused(p_2, s_2)$, $G = caused(q, s_1) \lor caused(q, s_2)$ is derived in SLC, meaning that s_1 and s_2 jointly cause q. The corresponding inference can be obtained in GLC as follows.

$$\cfrac{\cfrac{\cfrac{s_1 \vDash p_1}{s_1 \land s_2 \vDash p_1} \quad \cfrac{s_2 \vDash p_2}{s_1 \land s_2 \vDash p_2} \; (\text{Strengthening})}{s_1 \land s_2 \vDash p_1 \land p_2} \; (\text{And}) \quad \cfrac{p_1 \land p_2 \mapsto q}{p_1 \land p_2 \vDash q} \; (\text{Reflection})}{s_1 \land s_2 \vDash q} \; (*)$$

The last inference (*) is by Transitivity, which is obtained by Strengthening and Cut. This proves that $G' = (s_1 \land s_2 \vDash q)$ holds. $\qquad \square$

---

[15]One would claim that an abductive consequence relation should not be the same as that for deduction, so that some restriction is put in the condition of a rule to work with it under an abductive context [18, 48].

[16]In input/output logics [43], the motivation and rationale of inference rules (9–12,14) are explained from the philosophy of norms and deontic logic [44]. Although norms and causal statements are quite different in nature, Bochman adopted a similar set of rules for his logic of causality [5] and a causal logic of abduction [6].

Let $\preceq$ be a consequence relation such that $\alpha \preceq \beta$ iff $\beta \succeq \alpha$.[17] Writing down the meta-rules (8–12,14) in a logic program at the meta-level, we can define MLA on GLC.

Definition 3.4. A theory of *General Logic of Causality* (GLC) is represented as a meta-theory T_{GLC} that consists of a set of facts with the meta-predicate *linked* mentioning object-level causal links, together with the meta-rules:

$$caused(X, Y) \leftarrow linked(X, Y). \tag{15}$$
$$caused(X, Y) \leftarrow (X \preceq Z) \wedge caused(Z, Y). \tag{16}$$
$$caused(X, Y) \leftarrow caused(X, Z) \wedge (Z \preceq Y). \tag{17}$$
$$caused(X \wedge Y, Z) \leftarrow caused(X, Z) \wedge caused(Y, Z). \tag{18}$$
$$caused(X, Y) \leftarrow caused(X, Y \wedge Z) \wedge caused(Z, Y). \tag{19}$$
$$caused(X, Y \vee Z) \leftarrow caused(X, Y) \wedge caused(X, Z). \tag{20}$$

A theory for *GLC-based MLA* is an MLA framework $(T_{GLC}, \mathcal{A}_{GLC}, G)$, in which $\mathcal{A}_{GLC}$ is a set of facts with the meta-predicate *linked* and G is a fact with the meta-predicate *caused*.

In the same way as SLC-based MLA, law abduction with predicate invention is realized in GLC-based MLA. A conjunctive cause can be abduced in the form of $linked(g, s \wedge t)$ in GLC rather than a disjunction of *linked* literals in SLC. Compared with SLC, GLC is more complete but generally needs more inference steps for an underlying abductive procedure, since there are more possible rule instances to be examined in each step so that the possible search space for GLC is larger than that for SLC.

As mentioned earlier, GLC does not provide a specific correspondence between the meta-level and object-level formulas. Instead, the correctness of MLA is given with respect to the inference system.

Proposition 3.0.3. *Suppose a GLC-based MLA $(T_{GLC}, \mathcal{A}_{GLC}, G)$. Let E be a conjunction of atoms from $\mathcal{A}_{GLC}$. Then, E is an explanation of $G = caused(g, s)$ if and only if $(s \vDash g)$ is derived from $T_{GLC} \wedge E$ using the inference rules (8–12,14).*

Proof. Given a causal network N, we can show that $G = caused(g, s)$ is derived from N and the rules (15–20) if and only if $(s \vDash g)$ is derived from N using the inference rules (8–12,14). Let T interchangeably denote both a logic program defined with N and the rules (15–20) and the inference system with N and the rules (8–12,14). The proposition then easily follows by replacing T with $T_{GLC} \wedge E$. $\qquad\square$

---

[17]If α and β are conjunctions of propositional atoms, $\alpha \preceq \beta$ is equivalent to $\alpha \subseteq \beta$, where a conjunction is identified with the set of its conjuncts.

In summary, GLC offers a logic for causal networks in a general setting. Unlike SLC, GLC does not need a translation of meta-level formulas into object-level formulas. Like SLC, GLC-based MLA realizes 2nd-order existential abduction at the object level.

3.3 Causality Logic with Inhibition

In GLC-based MLA, the meaning of causes was not explicitly argued, yet links in a causal network have been of one kind, i.e., $linked(g, s)$, representing that g directly depends on s somehow. Depending on applications, GLC can give a specific semantics to the relation $linked$; but once its semantics is determined, it is applied to all causal statements. However, this is not desirable in many real-world situations, since there are multiple types of causalities in general. At least, we can easily consider two types of causalities, i.e., *positive* and *negative* causal effects. With this regard, we can understand that each arc of the form $linked(g, s)$ only represents positive effects.

In [31], applicability of MLA is thus extended to deal with networks containing both positive and negative causal effects. Such networks are seen in many domains like physics, ecology, economics and social science. In particular in biological domains, *inhibition* effects negatively in gene regulatory, signaling and metabolic pathways. Two types of direct causal relations, *triggered* and *inhibited*, are thus introduced as meta-predicates in [31]. Note that we still do not give any specific meaning to positive and negative causality but just require that they should be opposed notions. The notion of causal chains is also divided into two types: the positive one (*promoted*) and the negative one (*suppressed*), respectively corresponding to chained relations of *triggered* and *inhibited*.

Definition 3.5. A theory of *Logic of Causality with Inhibition* (LCI) is represented as a meta-theory T_{LCI} that consists of a set of (disjunctive) facts with the meta-predicates *triggered* and *inhibited* mentioning object-level causal relations, together with the meta-rules:

$$promoted(X, Y) \leftarrow triggered(X, Y). \tag{21}$$

$$promoted(X, Y) \leftarrow triggered(X, Z) \wedge promoted(Z, Y). \tag{22}$$

$$promoted(X, Y) \leftarrow inhibited(X, Z) \wedge suppressed(Z, Y). \tag{23}$$

$$suppressed(X, Y) \leftarrow inhibited(X, Y). \tag{24}$$

$$suppressed(X, Y) \leftarrow inhibited(X, Z) \wedge promoted(Z, Y). \tag{25}$$

$$suppressed(X, Y) \leftarrow triggered(X, Z) \wedge suppressed(Z, Y). \tag{26}$$

A theory for *LCI-based MLA* is an MLA framework $(T_{LCI}, \mathcal{A}_{LCI}, G)$, in which $\mathcal{A}_{LCI}$ is a set of (disjunctive) facts with the meta-predicates *triggered* and *inhibited* and G is a (disjunctive) fact with the meta-predicate *promoted* and *suppressed*.

LCI-based MLA reduces to SLC-based MLA when there is no inhibitor in a causal network. In this case, the first two axioms (21) and (22) for *promoted* have the same structure as (5) and (6). In the presence of inhibitors, we assume that a node X can be promoted if an adjacent inhibitor for X is blocked (23). On the other hand, a negative causal chain to X can be established if negative influence is propagated to X either directly by an adjacent inhibitor of an active (24) or promoted (25) item or indirectly by a trigger of suppressed (26). In this way, we can establish the connection between X and Y by mixing both positive and negative links. By (21–26), all possible paths from a source to a target, which is either positive or negative, can be obtained by meta-level abduction.

In this formalization of LCI-based MLA, both $promoted(g, s)$ and $suppressed(g, s)$ can be explained at the same time. This is not surprising because we can give any virtual goal and abduction can achieve it either positively or negatively. Although different sets of abducibles lead to different consequences, incompatible conclusions can be derived from the same set of abducibles. To detect such a potential inconsistency of a combination of abducibles, the following integrity constraint can be placed at the meta-level.

$$\leftarrow promoted(X, Y) \wedge suppressed(X, Y). \tag{27}$$

Incorporation of negative effects in causal networks introduces non-linear dynamics into the system. In the logical setting, this phenomenon is related to *nonmonotonic reasoning*. For example, biological networks are modeled to prefer inhibition to activation in [31]: a trigger of g works if there is no inhibitor for g, but if an inhibitor is added then the trigger stops working. MLA has been combined with default reasoning in [31] to implement this nonmonotonicity in the framework of abduction with *default assumptions* [33]. Alternatively, this nonmonotonic version can be expressed by replacing the three rules (21,22,26) containing triggers with normal defaults in *default logic* [66]:

$$\frac{triggered(X, Y) \ : \ promoted(X, Y)}{promoted(X, Y)},$$

$$\frac{triggered(X, Z) \wedge promoted(Z, Y) \ : \ promoted(X, Y)}{promoted(X, Y)},$$

$$\frac{triggered(X, Z) \wedge suppressed(Z, Y) \ : \ suppressed(X, Y)}{suppressed(X, Y)}.$$

Applications to p53 signal networks [64] are presented in [31] as case studies, in which meta-level abduction reproduces theories explaining how tumor suppressors work [75]. Analysis of such abstract signaling networks, although simple, tackles a fundamental inference problem in computer-aided scientific research including Systems Biology, such that, given an incomplete causal network, the goal is to infer possible connections and functions of network entities in order to reach the target entities from the sources.[18] Network

---

[18] Several logic-based models of biological systems and reasoning on them can be found in [16].

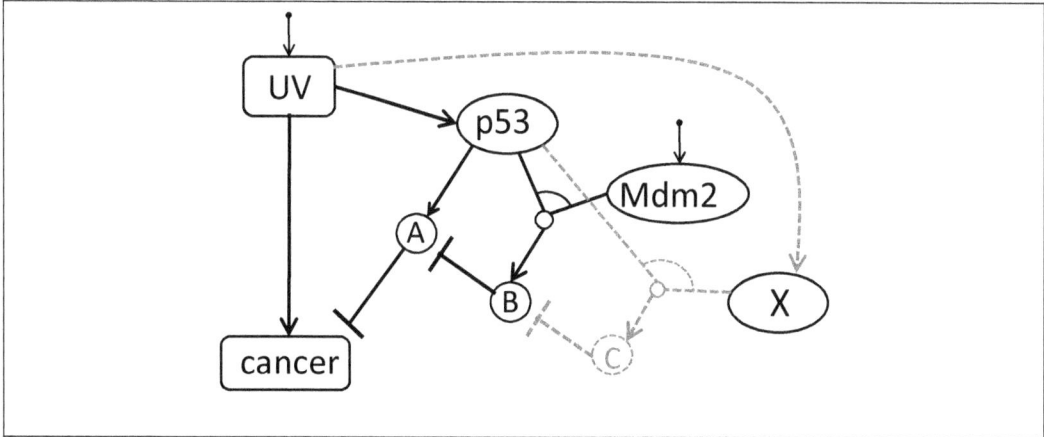

Figure 1: Completion of the p53 network [75, 31]

completion in signaling networks is recognized particularly important, since it is hard to observe activity levels and quantities of proteins in living organisms [1]. This feature of non-observability also exists in completing causal networks with respect to physical skills for cello playing [32], in which internal causation in human bodies cannot be observed directly. Hence abduction in these domains is inevitably *unobservable fact/law abduction*. Moreover, it is *micro-part abduction* or *hypothetical cause abduction* defined by Schurz [71], both of which belong to 2nd-order existential abduction.

Inoue *et al.* [31] show how the nonmonotonic version of LCI-based MLA works for the following biological scenario made by [75].

> Ultraviolet light (UV) causes stress, which influences the growth of tumors and induces the upregulation of the p53 protein. p53 leads to protect the target cell from cancer by binding its transactivator domain to the promoters of the gene. But the activity of p53 in the cell is influenced by its interactions with other proteins. The mediator complex p53/Mdm2, which is formed by binding the transactivator domain of p53 to Mdm2, inhibits p53 from tumor suppression.

This scenario can be simply represented at the meta-level by the facts:

$$triggered(\text{cancer}, \text{uv}), \quad triggered(\text{p53}, \text{uv}),$$
$$inhibited(\text{cancer}, a), \quad triggered(a, \text{p53}), \tag{28}$$
$$inhibited(a, b), \quad triggered(b, \text{p53} \wedge \text{mdm2}),$$

where "a" is the inhibitory domain of p53 and "b" is the p53/Mdm2 complex. In [31], the atom of the form $triggered(X, Y \wedge Z)$ is replaced with the disjunction ($triggered(X, Y) \vee$

25

$triggered(X, Z))$ like (4) in SLC.[19] Alternatively, we can introduce meta-axioms to deal with a conjunction in an argument of $promoted/2$ and $suppressed/2$ like (16–19) in GLC.[20] The scenario is depicted in the solid lines of the causal network in Figure 1. Note that effect of p53 in the target cell is nonmonotonic. If only UV exists, cancer may appear. But then p53 is activated by UV, which suppresses cancer. Then, if Mdm2 is introduced and is bound with p53, cancer appears again. That is, inhibiting an inhibition (*double inhibitions*) leads to promotion.

In the setting of [75], a biologist considers some another tumor suppressor gene X, whose function is not well known but whose mutants are highly susceptible to cancer. In an experiment, exposure of the cell to high level UV does not lead to cancer, under the conditions that the initial concentration of Mdm2 is high and that a high level of gene expression of the X protein is also observed. Hence we are given an empirical rule

$$\exists S(suppressed(\texttt{cancer}, S)),$$

which should be achieved by MLA with the abducibles $\{triggered(\_,\_), inhibited(\_,\_)\}$ and by finding a source S of this suppression. The goal thus cannot be given as a simple input-output pair, but should contain an unknown source S, since it is hard to identify which conditions suppress cancer.

Then LCI-based MLA produces the following hypothesis among others:

$$triggered(\texttt{x}, \texttt{uv}) \wedge \exists Y (triggered(Y, \texttt{p53} \wedge \texttt{x}) \wedge inhibited(b, Y)). \tag{29}$$

This hypothesis is depicted in dashed lines in Figure 1. The variable Y in (29) represents a new complex synthesized from X and p53. Hence, MLA generates a new mutant Y ("C" in Figure 1) by combining X and p53, which then inhibits the existing complex b. The hypothesis (29) indicates that X influences p53 protein stability. UV causes stress then induces high expression of X, which then binds to p53, so p53 is stabilized and formation of Mdm2-p53 complex is prevented. Hence, p53 ("A") can be functional as a tumor suppressor. This solution inverts the activation effect of double inhibitions again, that is, the triple inhibitions finally result in the suppression. The hypothesis (29) is actually suggested in [75], but an important thing here is that all abduced links in such a hypothesis are newly generated by MLA and that the new node Y is automatically invented during inference.[21]

---

[19]In [31], the correspondence between meta-level atoms and object-level formulas are given as follows: A trigger $triggered(g, s)$ is interpreted as $(g \leftarrow s) \wedge (\neg g \leftarrow \neg s)$ at the object level, and an inhibitor $inhibited(g, s)$ is interpreted as $(\neg g \leftarrow s) \wedge (g \leftarrow \neg s)$ at the object level. It turns out that the implication $\leftarrow$ in this correspondence cannot be equivalent to material implication in propositional logic.

[20]This extension of GLC to deal with both positive and negative causal effects has not yet been completed, and its nonmonotonic version with default logic is left open too.

[21]In contrast, Tran and Baral [75] consider an action theory for this p53 network, and prepare all possible ground candidate nodes and links as abducibles in advance. Hence, [75] performs selective law abduction, in which no new concept is automatically generated.

Those obtained hypotheses are important in the sense that the activation/inhibition mechanism of p53 is linked to some proteins that might not have been found out yet. MLA is thus crucial for this discovery task, and inferred hypotheses can suggest to scientists necessary experiments with gene knockout mice as minimally as possible.

In summary, LGI-based MLA can realize 2nd-order existential abduction, can abduce unobservable facts, new rules and new concepts, and can invent new objects and relations at the object level. Hence, LGI-based MLA is creative abduction.

4 Law/Theoretical-Model Meta-Level Abduction

Definition 4.1. A theory for *law MLA* is an MLA framework $(T, \mathcal{A}, G)$, in which $\mathcal{A}$ is a set of rules from the language $\mathcal{L}$.

The difference between fact MLA and law MLA lies only in the form of their explanations. As we have already mentioned in Section 2, selective law abduction can be simulated by fact abduction by the naming method. Selective law MLA is hence reduced to fact MLA in Section 3. On the other hand, theoretical-model abduction and law abduction are not necessarily distinguished from the point of view of computation, as discussed in Section 2.1. Hence, we consider *creative law MLA* here.

Law MLA could be called *meta-level induction*, since abduction of rules, i.e., explanatory induction is performed at the meta-level. Theoretically, any ILP system that is capable of law abduction can be used for law MLA. But, what can be abduced by law MLA? The main motivation of law MLA is to abduce meta-level axioms. Such meta-theories are found for deduction, nonmonotonic reasoning [37], abduction [18, 48, 42, 3, 6], causal theories [43, 5], deontic logic [44], etc. If we abduce laws for deduction by law MLA, it is abduction of deductive systems. If abduced laws are for abduction, it is abduction of abductive systems. Therefore, by law MLA, we can devise a part of deductive/abductive system.

One way to construct meta-rules for a domain by law MLA is to learn them from examples and a part of background meta-theory in the spirit of meta-level induction. For example, let T be a theory consisting of an incomplete meta-theory from T_{SLC} and the meta-expression of direct causal links:

$$T = (6) \land linked(a, b) \land linked(b, c).$$

Given $G = caused(a, c)$ as the observation, CF-induction [30] and some other ILP systems can construct the meta-rule (5) as a hypothesis to complete the meta-axioms of T_{SLC}. We can see that this abductive computation is nothing but an ordinary hypothesis-finding in ILP, but a meta-rule is abduced with respect to a meta-theory.

One might think that we can use *meta-meta-rules* to abduce meta-rules by fact MLA in the same way that meta-rules are used to abduce object-level rules by fact MLA. In this case, a meta-meta-interpreter is necessary, but often the same meta-interpreter is reused as a meta-meta-interpreter in logic programming. Now let us see the work by Bundy and Sterling [8], in which a meta-theory of algebra is proposed to derive new control information by proving theorems and a meta-meta-level language is developed to control the search for proving theorems by expressing proof plans. Here is an example of meta-meta-rules in [8]:

$$prove\text{-}by\text{-}induction(Conjecture) \leftarrow$$
$$structural\text{-}induction(Conjecture, Scheme) \land$$
$$prove\text{-}base\text{-}case(Conjecture, Scheme) \land$$
$$prove\text{-}step\text{-}case(Conjecture, Scheme).$$

Using this meta-meta-axiom and given a target equation $eq1$ and a partial inductive proof including the base case proof, abducing $prove\text{-}step\text{-}case(eq1, list\text{-}recursion)$ by fact MLA does not involve law abduction. Although this hypothesis tells us that the scheme of list recursion can be used for a proof plan, the meta-meta-atom alone does not provide a meta-theory on how to prove the step case but should be defined by another meta-meta-axiom. That is to say, we must distinguish meta-meta-level abduction from meta-level abduction of meta-theories, as in the case where meta-level abduction was distinguished from meta-abduction in Section 1.3.

5   Representation Meta-Level Abduction

Hoffmann's representation abduction or *meta-diagrammatic abduction* [27] requires representational changes for a theory. As its name indicates, meta-diagrammatic abduction itself is a kind of meta-level abduction that abduces a new domain theory. This implies that the underlying assumption of the language $\mathcal{L}$ of MLA considered in Section 3 such that the domain of $\mathcal{L}$ should contain all possible individuals might not be adequate. Hence, it is hard to realize representation abduction even at the object level, regardless of with or without MLA.[22] Consideration of representation MLA is thus done by a thought experiment.

A representation change often appears in scientific discovery as a paradigm shift when a totally new theory emerges. For example, non-Euclidean geometry changed geometrical postulates of Euclidean geometry, and classical physics was replaced by quantum physics with revolutionary ideas. In the course of such a discovery, a new representation system is abduced, and a new axiom set is introduced into the new framework. Hence abducing an essential change in a domain is representation abduction at the object level. On the other hand, for such invented non-Euclidean geometries, Hilbert studied axiomatic systems of

---

[22]Hoffmann [27] includes *selective meta-diagrammatic abduction* in his classification (see Table 2), which could be computed by MLA. However, it is not clear how a collection of representations is stored in advance.

both Euclidian and non-Euclidian geometries in his *"Foundations of Geometry"* [22]. As Hilbert later invoked the term *metamathematics*, his work on axiomatic systems exactly describes a meta-theory of those variants of geometries. Therefore, consideration of these axioms is exactly representation MLA.

In general, extraction of a hidden logic from a domain or inventing a new logic from observations in a new situation invokes representation MLA. This inference would also be necessary for AI programs or agents to automatically learn logics from observations. For this, Sakama and Inoue [70] consider the question "Can machines learn logics?":

> There are two machines $\mathcal{A}$ and $\mathcal{B}$. The machine $\mathcal{A}$ is capable of deductive reasoning: it has a set $\mathcal{L}$ of axioms and inference rules. Given a set S of formulas as an input, the machine $\mathcal{A}$ produces (a subset of) the logical consequences $Th(S)$ as an output. On the other hand, the machine $\mathcal{B}$ has no axiomatic system for deduction, while it is equipped with a machine learning algorithm $\mathcal{C}$. Given input-output pairs $(S_1, Th(S_1)), \ldots, (S_i, Th(S_i)), \ldots$ of $\mathcal{A}$ as an input to $\mathcal{B}$, the problem is whether one can develop an algorithm $\mathcal{C}$ which successfully reproduces the axiomatic system $\mathcal{L}$ for deduction.

Meta-level one-step deduction rules including MP are learned in [70]. Although this problem supposes learning of deduction, the scenario can be applied to learning abduction and other non-standard logics. The input examples to representation MLA is a set of observed input-output pairs in the form of pairs of input formulas and their consequences in the target logic. These input examples can be either given in such an abstract way as $(\{p \lor q, \ p \to r, \ q \to r)\}, \{p \lor q, \ r\})$ or can come from a specific domain like Euclidian geometry in the form of pairs of domain axioms and their consequences. Moreover, they are not necessarily given from a teacher, but can be implicitly hidden in log files of dynamic systems or in dialogues with unknown agents, so that the machine should identify those input-output relations automatically. Then the output of representation MLA is a set of meta-theoretical inference rules for the domain or inference patterns of the agents. To solve this problem, simple machine learning from observable samples would not work, but abduction of the underlying logic is necessary.[23] Once a logic is learned in this way, it must be refined or revised through a continuous, cyclic process between observations and abduction [19, 45] on meta-theoretical relations. This process would thus introduce a dynamics of incremental perfection of theories.

A similar mechanism is necessary for one to acquire or learn complex items like skills of music playing, cultural knowledge, science and advanced technologies, in which one

---

[23]There is an attempt to learn the underlying rules of dynamics from traces of state transition [34], which is applied to learn cellular automata and gene regulatory networks. This idea could be extended to learn the underlying logic in more general settings.

tries to explain them or understand what causes them with the intellectual faculties rather than by surface learning through experience.

6 Conclusion

The method of meta-level abduction proposed in [32] has been analyzed from the viewpoint of patterns of abduction. By generalizing the framework of MLA in [32], several patterns of MLA are considered in this paper. Fact MLA is strong enough to realize 2nd-order existential abduction at the meta-level. Law MLA can produce meta-axioms in logics. Representation MLA can abduce logics themselves.

To realize law abduction by fact MLA, we have mainly focused on the logic of causality, i.e., SLC and GLC, in the form of causal networks. Other meta-theories like control of inference [59, 8] can be used for meta-level abduction too. For such meta-programs, MLA will abduce a control atom like [60], but unlike SLC/GLC-based MLA this abduction is not law abduction any more. Hence, the expectation in Section 2.2 that MLA can lift the type of abduction to one level higher is not always true, but depends on the meta-theory. SLC and GLC deal with meta-theories for causal networks, and hence are useful for establishing and reconstructing causal relations between items. That is why fact abduction on these meta-theories can lead to law abduction at the object level.

There are several open problems on MLA. Technically, extension of GLC to deal with inhibition like LCI is a future work. Inhibition effects introduce negative effects, but there are other types of negations: an inhibition usually prevents causes, and a constraint stipulates that some causal statement never happens. We should distinguish those several meanings of negations in a logical way. For example, the statement that "α is not cause by β" can be expressed as $\neg caused(\alpha, \beta)$. Inference of negation could be implemented by either negation as failure or explicit negation in logic programming. Instead of using default logic in Section 3.3, we could express default reasoning in logic programs with negation.

So, why don't we use MLA more in various domains? In [31], it has been argued:

> Problem solving with meta-level abduction consists of three steps: (I) design of meta-level axioms; (II) representation of domain knowledge at the meta-level; and (III) restriction of the search space to treat large knowledge. This work supposes an incomplete network for Step (II), and hence representation of a problem is rather tractable. On the other hand, we have made great effort on Step (I), by gradually extending the positive causal networks to alternating axioms and then to nonmonotonic formalization. This formalization process was based on trial and error by running SOLAR many times.

Law MLA can thus contribute to Step (I) above. We did not discuss about Step (III) in this paper, and it is left open for both fact MLA and law MLA. This last topic is important not only for efficient computation of hypotheses but for one to select appropriate hypotheses easily. Discovering useful methods for constraint generation is left open, and another use of meta-level abduction such as control of reasoning will be useful too. Finally, refining theories through continuous uses of meta-theoretical abduction should be more investigated.

Acknowledgements

The author would like to thank reviewers for valuable comments on this paper, and is also indebted to Gauvain Bourgne, Andrew Cropper, Christine Froidevaux, Koichi Furukawa, Stephen Muggleton, Hidetomo Nabeshima, Adrien Rougny, Chiaki Sakama and Yoshitaka Yamamoto for fruitful discussions on the topic of this paper.

References

[1] Akutsu, T., Tamura, T. and Horimoto, K., Completing networks using observed data, *Proceedings of ALT '09*, LNAI 5809, pp.126–140. Springer (2009)

[2] Alferes, J.J., Pereira, L.M. and Swift, T., Abduction in well-founded semantics and generalized stable models via tabled dual programs, *Theory and Practice of logic Programming*, 4(4):383–428 (2004)

[3] Aliseda, A., *Abductive Reasoning: Logical Investigations into Discovery and Explanations*, Synthese Library, Vol.330. Springer (2006)

[4] Bharathan, V. and Josephson, J.R., Belief revision controlled by meta-abduction, *Logic Journal of the IGPL*, 14(2), 271–286 (2006)

[5] Bochman, A., A causal approach to nonmonotonic reasoning, *Artificial Intelligence*, 160:105–143 (2004)

[6] Bochman, A., A causal theory of abduction, *Journal of Logic and Computation*, 17(5), 851–869 (2007)

[7] Bowen, K.A. and Kowalski, R.A., Amalgamating language and metalanguage in logic programming, in: Clark, K. and Tarnlund, S. (eds.), *Logic Programming*, pp.153–172. Academic Press (1982)

[8] Bundy, A. and Sterling, L., Meta-level inference: Two applications, *Journal of Automated Reasoning*, 4(1), 15–27 (1988)

[9] Christiansen, H., Abduction and induction combined in a metalogic framework, in: [19], pp.195–211 (2000)

[10] Clark, K.L., Negation as failure, in: Gallaire, H. and Minker, J. (eds.), *Logic and Data Bases*, pp.119–140. Plenum Press, New York, 1978.

[11] Console, L., Dupre, D.T., and Torasso, P., On the relationship between abduction and deduction, *J. Logic and Computation*, 1(5), 661–690 (1991)

[12] Corapi, D., Russo, A. and Lupu, E.C., Inductive logic programming as abductive search, *Technical Communications of ICLP '10*, pp.54–63 (2010)

[13] Costantini, S., Meta-reasoning: A survey, in: Kakas, A.C. and Sadri, F. (eds.), *Computational Logic: Logic Programming and Beyond, Essays in Honour of Robert A. Kowalski, Part II*, LNAI 2408, pp.253–288. Springer (2002)

[14] Cox, P.T. and Pietrzykowski, T., Causes for events: Their computation and applications, in: *Proceedings of CADE '86*, LNCS 230, pp.608–621. Springer (1986)

[15] Eshghi, K., Abductive planning with event calculus, in: *Proceedings of the 5th International Conference on Logic Programming*, pp.562–579 (1988).

[16] Fariñas del Cerro, L. and Inoue, K. (eds.), *Logical Modeling of Biological Systems*. Wiley-ISTE (2014).

[17] Flach, P.A., *Simply Logical—Intelligent Reasoning by Example*. John Wiley (1994)

[18] Flach, P.A., Rationality postulates for induction, in: *Proceedings of TARK '96*, pp.267–281 (1996)

[19] Flach, P.A. and Kakas A.C. (eds.), *Abduction and Induction—Essays on their Relation and Integration*. Kluwer Academic (2000)

[20] Furukawa, K., Kinjo, K., Ozaki, T. and Haraguchi, M., On skill acquisition support by analogical rule abduction, *Revised Selected Papers from ISIP 2013*, CCIS 421, pp.71–83. Springer (2014)

[21] Gabbay, D. M. and Woods, J., *The Reach of Abduction: Insight and Trial*. Elsevier Science (2005)

[22] Hilbert, D., *The Foundations of Geometry*, English translation of *Grundlagen der Geometrie* (1902) by Townsend, E.J. (2nd ed.). Open Court Publishing, La Salle, IL (1950)

[23] Hill, P.M. and Gallagher, J., Meta-programming in logic programming, in: Gabbay, D.M., Hogger, C.J. and Robinson, J.A. (eds.), *Handbook of Logic in Artificial Intelligence and Logic Programming*, Vol.5, pp.421–497. Oxford University Press (1998)

[24] Hill, P.M. and Lloyd, J.W., Analysis of metaprograms, in: Abramson, H. and Rogers, M.H. (eds.), *Meta-Programming in Logic Programming*, pp.23–51. MIT Press (1988)

[25] Hitchcock, C., The intransitivity of causation revealed in equations and graphs, *The Journal of Philosophy*, 98(6):273–299 (2001)

[26] Hirata, K., A classification of abduction: Abduction for logic programming, *Machine Intelligence*, 14, 397–424 (1995)

[27] Hoffmann, M.H.G., "Theoric transformations" and a new classification of abductive inferences, *Transactions of the Charles S. Peirce Society*, 46(4), 570–590 (2010)

[28] Inoue, K., Linear resolution for consequence finding, *Artificial Intelligence*, 56, 301–353 (1992)

[29] Inoue, K., Automated abduction, in: Kakas, A.C. and Sadri, F. (eds.), *Computational Logic: Logic Programming and Beyond, Essays in Honour of Robert A. Kowalski, Part II*, LNAI 2408, pp.311–341. Springer (2002)

[30] Inoue, K., Induction as consequence finding, *Machine Learning*, 55, 109–135 (2004)

[31] Inoue, K., Doncescu, A. and Nabeshima, H., Completing causal networks by meta-level abduction, *Machine Learning*, 91(2), 239–277 (2013)

[32] Inoue, K., Furukawa, K., Kobayashi, I. and Nabeshima, H., Discovering rules by meta-level abduction, in: De Raedt, L. (ed.), *Inductive Logic Programming: Revised Papers from the 19th International Conference (ILP '09)*, LNAI 5989, pp.49–64. Springer (2010)

[33] Inoue, K., Iwanuma, K. and Nabeshima, H., Consequence finding and computing answers with defaults, *Journal of Intelligent Information Systems*, 26, 41–58 (2006)

[34] Inoue, K., Ribeiro, T. and Sakama, C., Learning from interpretation transition, *Machine Learning*, 94(1), 51–79 (2014)

[35] Josephson, J.J. and Josephson, S.G., *Abductive Inference: Computation, Philosophy, Technology*, Cambridge University Press (1994)

[36] Kakas, A.C., Kowalski, R.A. and Toni, F., The role of abduction in logic programming, in: Gabbay, D.M., Hogger, C.J. and Robinson, J.A. (eds.), *Handbook of Logic in Artificial Intelligence and Logic Programming*, Vol.5, pp.235–324. Oxford University Press (1998)

[37] Kraus, S., Lehmann, D. and Magidor, M., Nonmonotonic reasoning, preferential models and cumulative logics, *Artificial Intelligence*, 44, 167–207 (1990)

[38] Kowalski, R.A., Problems and promises of computational logic, in: Lloyd, J.W. (ed.), *Computational Logic*, pp.1–36 (1990)

[39] Kowalski, R.A. and Sergot, M.J., A logic-based calculus of events, *New Generation Computing*, 4(1), 67–95 (1986)

[40] Lee, C.T., *A completeness theorem and computer program for finding theorems derivable from given axioms*, Ph.D. thesis, Department of Electrical Engineering and Computer Science, University of California, Berkeley, CA (1967)

[41] Lejay, G., Inoue, K. and Doncescu, A., Application of meta-level abduction for the treatment of hypertension using SOLAR, *Proceedings of the AINA-2011 Workshop on Bioinformatics and Life Science Modeling and Computing*, pp.495–500. IEEE Computer Society (2011)

[42] Lobo, J. and Uzcátegui, C., Abductive consequence relations, *Artificial Intelligence*, 89, 149–171 (1997)

[43] Makinson, D. and van der Torre, L., Input/output logics, *Journal of Philosophical Logic*, 29, 383–408 (2000)

[44] Makinson, D. and van der Torre, L., What is input/output logics?, in: *Foundations of the Formal Sciences II: Applications of Mathematical Logic in Philosophy and Linguistics*, pp.163–174. Kluwer Academic (2003)

[45] Magnani, L., *Abduction, Reason, and Science: Processes of Discovery and Explanation*. Kluwer Academic/Plenum Publishers (2001)

[46] Magnani, L., *Abductive Cognition—The Epistemological and Eco-Cognitive Dimensions of Hypothetical Reasoning*, Cognitive Systems Monographs 3. Springer (2009)

[47] Marquis, P. Consequence finding algorithms, in: Gabbay, D.M. and Smets, P. (eds.), *Handbook for Defeasible Reasoning and Uncertain Management Systems*, Vol.5, pp.41–145. Kluwer Academic (2000)

[48] Mayer, M.C. and Pirri, F., Abduction is not deduction-in-reverse, *Logic Journal of the IGPL*, 4(1), 95–108 (2006)

[49] Muggleton, S. and Bryant, C., Theory completion and inverse entailment, in: *Proceedings of*

ILP 2000, LNAI 1866, pp.130–146. Springer (2000)

[50] Muggleton, S. and Buntine, W., Machine invention of first-order predicate by inverting resolution, in: *Proceedings of the 5th International Workshop on Machine Learning*, pp.339–351. Morgan Kaufmann (1988)

[51] Muggleton, S.H., De Raedt, L., Poole, D., Bratko, I., Flach, P., Inoue, K. and Srinivasan, A., ILP turns 20—Biography and future challenges. *Machine Learning*, 86(1), 3–23 (2012)

[52] Muggleton, S.H. and Lin, D., Meta-Interpretive learning of higher-order dyadic Datalog: Predicate invention revisited, *Proceedings of IJCAI-13*, pp.1551–1557 (2013)

[53] Muggleton, S.H., Lin, D., Pahlavi, N. and Tamaddoni-Nezhad, A., Meta-interpretive learning: Application to grammatical inference, *Machine Learning*, 94(1), 25–49 (2014)

[54] Nabeshima, H., Iwanuma, K. and Inoue, K., SOLAR: A consequence finding system for advanced reasoning, in: *Proceedings of TABLEAUX '03*, LNAI 2796, pp.257–263. Springer (2003)

[55] Nabeshima, H., Iwanuma, K., Inoue, K. and Ray, O., SOLAR: An automated deduction system for consequence finding, *AI Communications*, 23(2,3), 183–203 (2010)

[56] Paul, G., AI approaches to abduction, in: *Handbook of Defeasible Reasoning and Uncertainty Management Systems*, Vol.4, pp.35–98. Kluwer (2000)

[57] Pearl, J., *Causality: Models, Reasoning, and Inference*, 2nd Ed. Cambridge (2009)

[58] Peirce, C.S., *The Collected Papers*, Hartshorne, C. and Weiss, P. (eds.). Harvard University Press, Cambridge, MA (1935)

[59] Pereira, L.M., Logic control with logic, in: Campbell, J. (ed.), *Implementations of Prolog*, pp.177–193. Ellis Horwood (1984)

[60] Pereira, L.M. and Pinto, A.M., Inspecting side-effects of abduction in logic programs, in: *Logic Programming, Knowledge Representation, and Nonmonotonic Reasoning: Essays Dedicated to Michael Gelfond on the Occasion of His 65th Birthday*, LNAI, Vol.6565, pp.148–163. Springer (2011)

[61] Poole, D., A logical framework for default reasoning, *Artificial Intelligence*, 36, 27–47 (1988)

[62] Poole, D., Goebel, R. and Aleliunas, R., Theorist: A logical reasoning system for defaults and diagnosis, in: Cercone, N. and McCalla, G. (eds.), *The Knowledge Frontier: Essays in the Representation of Knowledge*, pp.331–352. Springer (1987)

[63] Pople, Jr, H.E., On the mechanization of abductive logic, In: *Proceedings of IJCAI-73*, pp.147–152 (1973)

[64] Prives, C and Hall, P.A., The p53 pathway, *Journal of Pathology*, 187, 112–126 (1999)

[65] Ray, O. and Inoue, K., A consequence finding approach for full clausal abduction, in: *Proceedings of the 10th International Conference on Discovery Science*, LNAI 4755, pp.173–184. Springer (2007)

[66] Reiter, R., A logic for default reasoning, *Artificial Intelligence*, 13, 81–132 (1980)

[67] Robinson, J.A., A machine-oriented logic based on the resolution principle, *Journal of the ACM*, 12, 23–41 (1965)

[68] Russell, S. and Wefald, E., Principles of metareasoning, *Artificial Intelligence*, 49, 361–395

(1991)

[69] Sakama, C., Abductive generalization and specialization, in: [19], pp.253–265 (2000)

[70] Sakama, C. and Inoue, K., Can machines learn logics?, *Proceedings of the 8th International Conference on Artificial General Intelligence*, LNAI 9205, pp.341–351, Springer (2015)

[71] Schurz, G., Patterns of abduction, *Synthese*, 164(2), 201–234 (2008)

[72] Shanahan, M., An abductive event calculus planner, *Journal of Logic Programming*, 44, 207–239 (2000)

[73] Shapiro, E.Y., Inductive inference of theories from facts, Research Report 192, Yale University (1981)

[74] Stickel, M.E., Upside-down meta-interpretation of the model elimination theorem-proving procedure for deduction and abduction, *Journal of Automated Reasoning*, 13(2), 189–210 (1994)

[75] Tran, N. and Baral, C., Hypothesizing about signaling networks, *Journal of Applied Logic*, 7(3), 253–274 (2009)

[76] Woodward, J., *Making Things Happen—A Theory of Causal Explanation*. Oxford University Press (2003)

[77] Yamamoto, Y., Rougny, A., Nabeshima, H., Inoue, K., Moriya, H., Froidevaux, C. and Iwanuma, K., Completing SBGN-AF networks by logic-based hypothesis finding, *Formal Methods in Macro Biology*, LNBI 8738, pp.165–179. Springer (2014)

Received January 2015

Abduction and Beyond
in Logic Programming
with Application to Morality

Luís Moniz Pereira

NOVA Laboratory for Computer Science and Informatics
Departamento de Informática, Faculdade de Ciências e Tecnologia
Universidade Nova de Lisboa, Portugal
lmp@fct.unl.pt

Ari Saptawijaya

Faculty of Computer Science at Universitas Indonesia, Depok, Indonesia
saptawijaya@cs.ui.ac.id

Abstract

In this paper we emphasize two different aspects of abduction in Logic Programming (LP): (1) the engineering of LP abduction systems, and (2) application of LP abduction, complemented with other non-monotonic features, to model morality issues. For the LP engineering part, we present an implemented tabled abduction technique in order to reuse priorly obtained (and tabled) abductive solutions, from one abductive context to another. Aiming at the interplay between LP abduction and other LP non-monotonic reasoning, this tabled abduction technique is combined with our own-developed LP updating mechanism – the latter also employs tabling mechanisms, notably incremental tabling of XSB Prolog. The first contribution of this paper is therefore a survey of our tabled abduction and updating techniques, plus further development of our preliminary approach to combine these two techniques.

The second contribution of the paper pertains to the application part. We formulate a LP-based counterfactual reasoning, based on Pearl's structural theory, via the aforementioned unified approach of our LP abduction and updating.

We thank the referees for their constructive comments and suggestions. Both authors acknowledge the support from FCT/MEC NOVA LINCS PEst UID/CEC/04516/2013. Ari Saptawijaya acknowledges the support from FCT/MEC grant SFRH/BD/72795/2010. We thank Terrance Swift and David Warren for their expert advice in dealing with implementation issues in XSB, and Emmanuelle-Anna Dietz for fruitful discussions on counterfactuals.

The formulation of counterfactuals allows us to subsequently demonstrate its applications to model moral permissibility, according to the Doctrines of Double and Triple Effect, and to provide its justification. The applications are shown through classic moral examples from the literature, and tested in our prototype, QUALM, an implementation reifying the presented unified approach.

1 Introduction

Abduction has been well studied in Logic Programming (LP) [11, 26, 25, 13, 17, 4], establishing theoretical results that support the implementation of LP abduction systems. LP abduction has been applied in a variety of areas, such as in diagnosis [18], planning [15], scheduling [28], reasoning of rational agents and decision making [31, 42], knowledge assimilation [27], natural language understanding [5], security protocols verification [1], and systems biology [50]. Such applications demonstrate the potential of LP abduction and motivates further advancements of LP abduction systems to prepare them for even more challenging applications.

In this paper we report two results of our recent research in LP abduction concerning: (1) the engineering of LP abduction systems, and (2) application of LP abduction, complemented with other non-monotonic reasoning, to modeling a number of issues in morality.

Engineering of LP abduction systems We address two challenges in engineering a LP abduction system. For one, in abduction, finding some best explanations (i.e. adequate abductive solutions) to an observed evidence, or finding assumptions that can justify a goal, can be costly. It is often the case that abductive solutions found within one context are also relevant in a different context, and can be reused with little cost. In LP, absent of abduction, goal solution reuse is commonly addressed by employing a tabling mechanism [62]. Therefore, tabling appears to be conceptually suitable for abduction, so as to reuse abductive solutions. In practice, abductive solutions reuse is not immediately amenable to tabling, because such solutions go together with an abductive context.

We conceptualize in [57] a *tabled abduction* technique, to benefit from LP tabling mechanisms in contextual abduction, i.e., to reuse priorly obtained (and tabled) abductive solutions, from one abductive context to another. The technique is underpinned by the theory of ABDUAL [4], for computing abduction in LP over Well-Founded Semantics [66]. The tabled abduction technique is implemented in XSB Prolog [64], resulting in a LP abduction system TABDUAL. It concretely realizes the abstract theory of ABDUAL, while also taking care of pragmatic issues, due to the presence of tabled abduction, to foster the practicality of TABDUAL.

Another challenge we address is the interplay between LP abduction and other LP non-monotonic reasoning. It is apparent that when it comes to applications, notably in reasoning of rational agents and decision making, abduction needs to be enriched with other features. For instance, such applications are susceptible to knowledge updates and changes, due to incomplete information of the world or the evolution of the agents themselves. In this case, abduction systems need to be complemented with LP updating feature.

Aiming at that goal, we propound yet another use of tabling mechanisms, notably the so-called *incremental tabling*, for LP updating. Incremental tabling, also available in XSB Prolog, is an advanced tabling feature that ensures the consistency of answers in a table with all dynamic information on which the table depends. It does so by incrementally maintaining the table, rather than by recomputing answers in the table from scratch to keep it updated. This incremental table maintenance is achieved by automatically propagating assertion or retraction of information bottom-up; such propagation is typically triggered by the invocation of a given goal. In [56, 55], we conceptualize a technique with incremental tabling that permits a reconciliation of high-level top-down deliberative reasoning about a goal, with autonomous low-level bottom-up world reactivity to ongoing updates. The technique, dubbed EVOLP/R, is theoretically based on Dynamic Logic Programs [3] and its subsequent development, Evolving Logic Programs (EVOLP) [2].

In [58] we preliminarily bring LP abduction and LP updating into a unified approach based on the aforementioned two implementation techniques: TABDUAL and EVOLP/R, respectively. It aims at the use of joint tabling of both features, so as to benefit from reusing the abductive solutions obtained in one context for another, while also allowing top-down goal-driven incremental tabling of updates by upwards propagation. The approach in [58] borrows the concept of hypothesis generation [42]. While this concept permits generating only abductive explanations relevant for the problem at hand, the preliminary unified approach [58] limits the benefit of tabled abduction. In this paper we further develop this unified approach by leaving out the hypotheses generation concept in order to uphold the idea and the benefit of tabled abduction. The unified approach is implemented and applied to formulate counterfactual reasoning, which in turn is used to address morality issues, as we detail next.

Applications of LP abduction to Morality Computational morality (also known as machine ethics, machine morality, artificial morality and computational ethics) is a burgeoning field of inquiry that emerges from the need of imbuing autonomous agents with the capacity of moral decision-making. It has particularly attracted interest from the artificial intelligence community and has brought together perspectives from various fields, amongst them: philosophy, cognitive science, neuroscience and primatology. The overall result of this interdisciplinary research is therefore not only important for equipping agents with the capacity of making moral judgments, but also for helping us better understand morality, through the creation and testing of computational models of ethical theories.

The potential of LP to computational morality has been reported in [30, 45, 21]. In [59] we summarize our work emphasizing the application of LP abduction and other LP-based reasoning for modeling the dynamics of knowledge and moral cognition of an individual agent, where we employ LP abduction jointly with other non-monotonic reasoning. The reader is also referred to [46] which summarizes: (1) our study with other co-authors on collective morals and the emergence, in a population, of evolutionarily stable moral norms, of fair and just cooperation, in the presence of cognitive aspects, such as intention recognition, commitment, and mutual tolerance through apology; and (2) our view on bridging individual and collective morality, as they are necessarily intertwined.

In the second part of the paper we report the continuation of our prior work for modeling the dynamics of knowledge and moral cognition of an individual agent, now benefiting from the integration of TABDUAL and EVOLP/R discussed earlier. It is important to note that the application we show in this paper neither aims at defending any considered moral principles nor resolving the dilemmas appearing in the examples, as even philosophers naturally are split over opinions on them. Instead, its purpose is to show that our unified approach of LP abduction and updating (with its implementation) are capable and appropriate for expressing viewpoints on morality issues discussed herein, in accordance with the results and the theory argued in the literature.

We start by applying LP abduction and updating to model counterfactual reasoning, as people typically reason about what they should or should not have done when they examine decisions in moral situations [36, 34, 52, 38]. It is therefore natural for them to engage counterfactual thoughts in such settings.

Counterfactuals capture the process of reasoning about a past event that did not occur, namely what would have happened had this event occurred; or, vice-versa, to reason about an event that did occur but what if it had not. An example, taken from [8]: *Lightning hits a forest and a devastating forest fire breaks out. The forest was dry after a long hot summer and many acres were destroyed.* One may think of a counterfactual about it, e.g., "if only there had not been lightning, then the forest fire would not have occurred".

Counterfactuals have been widely studied in philosophy [32, 9, 20], psychology [35, 51, 36, 8, 14, 38], as well as from the computational viewpoint [19, 41, 6, 40, 67]. Research in LP to model counterfactual reasoning is nevertheless still limited. In

[41], counterfactuals, based on Lewis's counterfactuals [32], are evaluated using contradiction removal semantics of LP. The semantics defines the most similar worlds by removing contradictions from the associated program, obtaining the so-called maximal non-contradictory submodels of the program. In [6], Probabilistic LP language P-log with Stable Model Semantics is employed to encode Pearl's probabilistic causal model [40] of counterfactual reasoning. In [67], Pearl's approach is encoded using a different Probabilistic LP, viz., CP-logic. None of these LP-based approaches involves abduction in modeling counterfactuals.

In this paper we specifically adopt Pearl's approach [40], abstaining from probabilities, but resorting to LP abduction and updating. LP abduction is employed for providing background conditions from observations made or evidences given, whereas defeasible LP rules allow adjusting the current model via hypothetical updates of intervention. Therefore, our first application is to formulate an implemented evaluation procedure for counterfactuals via LP abduction and updating. The implementation is based on our unified approach of TABDUAL and EVOLP/R, and named QUALM (its ongoing development is available from `https://github.com/merah-putih/qualm`). QUALM is appropriately used for applications concerned with the reasoning of agents, in particular those involving counterfactuals when probabilities are not known or needed. We specifically look into its challenging applications in moral reasoning (for our related work on probabilistic moral reasoning, cf. [21]).

We apply counterfactuals (by resorting to its LP formulation), together with a combination of LP abduction and updating, to model morality issues. Counterfactuals are engaged to examine moral permissibility of an action by distinguishing whether an effect of an action is a cause for achieving a morally dilemmatic goal or merely a side-effect of that action. The distinction is essential for establishing moral permissibility from the viewpoints of the Doctrines of Double Effect [37] and of Triple Effect [29], as scrutinized herein through several classic moral examples from the literature. We also apply counterfactuals to address a form of moral justification. Through examples, we show how compound counterfactuals may provide moral justification for what was done due to lack of the current knowledge. We also touch upon Scanlon's contractualism [60] as a reference for employing counterfactuals to provide a justified, but defeasible, exception to permissibility of actions.

The paper is organized as follows. Section 2 reviews necessary background on abduction in LP. We discuss the LP engineering part by presenting the concept of tabled abduction, LP updating with incremental tabling, and the interplay between them in Section 3. The application part is subsequently elaborated in the next two sections: the formulation of counterfactuals with LP abduction and updating is detailed in Section 4, whereas Section 5 presents applications of counterfactuals to model morality issues. The reader may skip the technical LP engineering part

(Section 3) without loss of understanding of the application part. We conclude in Section 6 with some remarks on related issues that may trigger future work.

2 Abduction in Logic Programming

We start by recapping basic notation in LP and review how abduction is expressed and computed in LP.

A *literal* is either an atom B or its default negation *not B*, named positive and negative literals, respectively. They are negation complements to each other. The atoms *true* and *false* are true and false, respectively, in every interpretation. A *logic program* is a set of rules of the form $H \leftarrow B$, naturally read as "H if B", where its head H is an atom and its (finite) *body* B is a sequence of literals. When B is empty (equal to *true*), the rule is called a *fact* and simply written H. A rule in the form of a denial, i.e., with *false* as head, is an *integrity constraint* (IC).

In LP, an abductive hypothesis (*abducible*) is a 2-valued positive literal Ab or its negation complement Ab^* (denotes *not Ab*), whose truth value is not initially assumed, and it does not appear in the head of a rule. An *abductive logic program* is one allowing abducibles in the body of rules. An observation O is a set of literals, analogous to a query in LP; we use the usual LP notation ?- to denote a query.

Abduction in LP can be accomplished by a top-down query-oriented procedure for finding a query solution, i.e., a *consistent* abductive solution, by need. The solution's abducibles are leaves in its procedural query-rooted call-graph, where the graph is recursively generated by the procedure calls from literals in bodies of rules to heads of rules, and thence to the literals in a rule's body.

Example 1. A library in a city is close with two possible explanations: it is close in the weekend, or when it is not weekend, there are no librarians working. These days, librarians are often absent from their work because they participate in a strike. On the other hand, a museum in that city is only close when there is a special visit by important guests. This example can be expressed by a logic program below (*weekend*, *strike*, and *special_visit* are abducible atoms):

$$close\_library \leftarrow weekend. \quad close\_library \leftarrow weekend^*, absent.$$
$$absent \leftarrow strike. \quad close\_museum \leftarrow special\_visit.$$

Consider the query ?- *close_library*. It induces the call-graph with *close_library* as its root and, through procedure calls, it ends with two leaves: one leaf containing abducible *weekend* as an abductive solution, and the other containing abducibles [*weekend**, *strike*] as another solution.

The correctness of this top-down computation requires the underlying semantics to be relevant, as it avoids computing a whole model (to warrant its existence) in finding an answer to a query. Instead, it suffices to use only the rules relevant to the query – those in its procedural call-graph – to find its truth value. In Example 1, rule *close_museum ← special_visit* is not relevant to answer the aforementioned query, and thus is not involved in the call-graph.

The 3-valued Well-Founded Semantics or WFS [66] enjoys this relevancy property [12], i.e., it permits finding only relevant abducibles and their truth value via the aforementioned top-down query-oriented procedure. Those abducibles not mentioned in the solution are indifferent to the query, e.g. abducible *special_visit* is indifferent to query ?- *close_library*. Due to its relevancy property, the approaches in this paper are based on WFS. Moreover, the XSB Prolog system [64], wherein all the prototypes of our research are implemented, computes the WFS.

3 Tabling in Contextual Abduction and LP Updating

The progress of LP has been promoting new techniques for engineering LP-based systems, most notably tabling mechanisms [62]. Tabling affords solutions reuse, rather than recomputing them, by keeping in tables subgoals and their answers obtained by query evaluation. It is supported, to a different extent, by a number of Prolog systems.

In abduction, finding explanations to an observed evidence, or finding assumptions that can justify a goal, can be costly. We discuss in Section 3.1, that abduction may benefit from tabling mechanisms by reusing priorly obtained abductive solutions from one abductive context to another. We subsequently discuss a unified approach of LP abduction and tabling via their joint tabling, in Section 3.2.

3.1 Tabling in Contextual Abduction

Motivation and Idea   Example 2 illustrates the motivation and the idea of tabled abduction.

Example 2. A plane may be hit by lightning, when it flies in a stormy area and at a low altitude. A case where it flies into a stormy area is when it enters cumulonimbus cloud (a dense towering cloud associated with thunderstorms). Being hit by lightning in such a stormy area may crash the plane. This example is expressed in the program below (*cumulonimbus* and *low_altitude* are abducible atoms):

$$storm ← cumulonimbus. \qquad hit\_lightning ← storm, low\_altitude.$$
$$crash ← hit\_lightning, storm.$$

Suppose three queries are invoked, asking for individual explanations of *storm*, *hit_lightning*, and *crash*, in that order. The first query, ?- *storm*, is satisfied simply by having [*cumulonimbus*] as the abductive solution for *storm*, and tabling it.

Executing the second query, ?- *hit_lightning*, amounts to satisfying the two subgoals in its body, i.e., abducing *low_altitude* followed by invoking *storm*. Since *storm* has previously been invoked, we can benefit from reusing its solution, instead of recomputing, given that the solution was tabled. That is, query ?- *hit_lightning* can be solved by extending the current ongoing abductive context [*low_altitude*] of subgoal *storm* with the already tabled abductive solution [*cumulonimbus*] of *storm*, yielding the abductive solution [*cumulonimbus*, *low_altitude*].

The final query ?- *crash* can be solved similarly. Invoking the first subgoal *hit_lightning* results in the priorly registered abductive solution: [*cumulonimbus*, *low_altitude*], which becomes the current abductive context of the second subgoal *storm*. Since [*cumulonimbus*, *low_altitude*] subsumes the previously obtained (and tabled) abductive solution [*cumulonimbus*] of *storm*, we can then safely take [*cumulonimbus*, *low_altitude*] as the abductive solution to query ?- *crash*.

Example 2 shows how a tabled abductive solution [*cumulonimbus*], the abductive solution of the first query ?- *storm*, can be reused from one abductive context of *storm* (viz., [*low_altitude*] in the second query, ?- *hit_lightning*) to another context (viz., [*cumulonimbus*, *low_altitude*] in the third query, ?- *crash*). In practice a repeatedly called rule, like *storm*, may have a body comprising many subgoals, causing potentially expensive recomputation of its abductive solutions, if they have not been tabled.

Contextual Abduction with Tabling Tabled abduction with its prototype TABDUAL consists of a program transformation that provides self-sufficient program transforms, which can be directly run to enact abduction by means of TABDUAL's library of reserved predicates. We describe below how the idea in Example 2 is conceptually realized by contextual abduction with tabling in TABDUAL. The reader is referred to [54] for a more formal and technical treatment of the transformation.

Example 2 indicates two key ingredients of tabled abduction in TABDUAL:

1. *Abductive contexts*, which relay the ongoing abductive solution from one subgoal to subsequent subgoals in the body of a rule, as well as from the head to the body of a rule, via *input* and *output* contexts.

2. *Tabled predicates*, which table the abductive solutions for predicates defined in the program, such that they can be reused from one abductive context to another.

Example 3 shows how these two ingredients figure in the program transform of Example 2.

Example 3. Consider rule $storm \leftarrow cumulonimbus$ of Example 2. A new predicate $storm_{ab}$ is introduced to table the abductive solution of $storm$,[1] defined as:

$$storm_{ab}([cumulonimbus]).$$

It is a simple tabled fact, as the rule of $storm$ is defined solely by the abducible $cumulonimbus$, without any other literals in its body.

We defined similar tabled predicates $hit\_lightning_{ab}$ and $crash_{ab}$ for the other two rules (of $hit\_lightning$ and $crash$, respectively):

$$hit\_lightning_{ab}(E) \quad \leftarrow \quad storm([low\_altitude], E). \tag{1}$$

$$crash_{ab}(E_2) \quad \leftarrow \quad hit\_lightning([\,], E_1), storm(E_1, E_2). \tag{2}$$

Predicate $hit\_lightning_{ab}(E)$ is a predicate that tables, in E, the abductive solution of $hit\_lightning$. Its definition follows from the original definition of $hit\_lightning$. Two extra (and new) parameters, that serve as *input* and *output contexts*, are added to the subgoal $storm$. Notice how $low\_altitude$, the abducible appearing in the body of the original rule of $hit\_lightning$, becomes the input abductive context of $storm$. This subgoal $storm$ requires a rule of $storm(I, O)$, defined in rule (3) below, that reuses the solution tabled in $storm_{ab}$ to produce the output abductive context O from its input context I.

Predicate $crash_{ab}$ tables the abductive solution for $crash$ in its argument E_2. The rule expresses that the tabled abductive solution E_2 of $crash_{ab}$ is obtained by relaying the ongoing abductive solution stored in context E_1 from subgoal $hit\_lightning$ to subgoal $storm$ in the body, given the empty input abductive context (i.e., [\,]) of $hit\_lightning$ (because there is no abducible by itself in the body of the original rule of $crash$).

Now that we have tabled predicates, the remaining rules to define are those that reuse the tabled solutions. That is, we define the rule of $storm(I, O)$ that reuses the solution tabled in $storm_{ab}$, and produce the output abductive context O from its input context I (the other rules $hit\_lightning(I, O)$ and $crash(I, O)$ are defined similarly):

$$storm(I, O) \leftarrow storm_{ab}(E), produce\_context(O, I, E). \tag{3}$$

This rule expresses that the output abductive solution O of $storm$ is obtained from the solution entry E of $storm_{ab}$ and the given input context I of $storm$, via the

---

[1]In XSB Prolog, this tabling effect is achieved by declaring predicate $storm_{ab}$ as a tabled predicate.

TABDUAL system predicate $produce\_context(O, I, E)$. This system predicate concerns itself with: whether E is already contained in I and, if not, whether there are any abducibles from E, consistent with I, that can be added to produce O. For example, $produce\_context$ in rule (1) adds the tabled solution $cumulonimbus$ to the input abductive context $[low\_altitude]$ of $storm$, thus producing the output $[cumulonimbus, low\_altitude]$. On the other hand, in rule (2), the tabled solution $cumulonimbus$ of $storm$ is already contained in the input abductive context $E_1 = [cumulonimbus, low\_altitude]$, yielding the latter as the output $E_2 = E_1$.

Note that if E is inconsistent with I then the specific entry E cannot be reused with I, $produce\_context$ fails and another entry E is sought. For instance, suppose $storm$ is invoked with the input abductive context $[cumulonimbus^*]$. This goal $storm([cumulonimbus^*], O)$ will fail, as $produce\_context$ ends up with an inconsistent solution containing $cumulonimbus$ and $cumulonimbus^*$. In summary, $produce\_context$ should guarantee that it produces a *consistent* output context O from I and E that encompasses both.

Abduction under Negative Goals   We have seen in Example 3 abduction performed under positive goals via abductive contexts and tabling. A common approach for abducing under negative goals is by first computing all abductive solutions of its corresponding positive goal, and then having to negate their disjunction. TABDUAL avoids such computation of all abductive solutions by employing the *dual transformation* [4] in contextual abduction.

The dual transformation is based on the idea of making negative goals 'positive' literals. It thus permits obtaining one abductive solution at a time, just as if we treat abduction under positive goals. The dual transformation defines for each atom A and its set of rules R in a program P, a set of dual rules whose head not_A is true if and only if A is false by R in the employed semantics of P. Note that, instead of having a negative goal $not\ A$ as the rules' head, we use its corresponding 'positive' literal, not_A. Example 4 illustrates the main idea of how the dual transformation is employed in TABDUAL.

Example 4. Recall Example 1 and the two rules of $close\_library$:

$$close\_library \leftarrow weekend. \qquad close\_library \leftarrow weekend^*, absent.$$

The dual transformation will create a set of dual rules for $close\_library$. These dual rules falsify $close\_library$ with respect to its two rules. That is, the rule of $not\_close\_library$ is defined by falsifying both the first rule *and* the second rule, denoted below by predicate $close\_library^{*1}$ and $close\_library^{*2}$, respectively:

$$not\_close\_library(C_0, C_2) \leftarrow close\_library^{*1}(C_0, C_1), \ close\_library^{*2}(C_1, C_2).$$

Note the input and output abductive context parameters that figure in the rule: C_0 and C_2, in the head, and similarly in each subgoal of the rule's body, where intermediate context C_1 relays the ongoing abductive solution from $close\_library^{*1}$ to $close\_library^{*2}$.

Next, we define how the first and the second rules of $close\_library$ are falsified, i.e., rules for $close\_library^{*1}$ and $close\_library^{*2}$. These rules are naturally defined by falsifying the body of $close\_library$'s first rule and second rule, respectively. In case of $close\_library^{*1}$: the first rule of $close\_library$ is falsified only by abducing the negation of $weekend$, viz., $weekend^*$. Therefore, we have:
$$close\_library^{*1}(I, O) \leftarrow weekend^*(I, O).$$

Notice that $weekend^*$ is abduced by invoking the subgoal $weekend^*(I, O)$. This subgoal is defined as follows (other abducibles, both positive and negative ones, transform similarly):

$$weekend^*(I, O) \leftarrow insert\_abducible(weekend^*, I, O). \tag{4}$$

where $insert\_abducible(A, I, O)$ is a TABDUAL system predicate that inserts abducible $weekend^*$ into input context I, resulting in output context O. As in the predicate $produce\_context$, it also maintains the consistency of the context, failing if inserting A results in an inconsistent one.

In case of $close\_library^{*2}$: the second rule of $close\_library$ is falsified by alternatively failing one subgoal in its body at a time, i.e. by negating $weekend^*$ or, alternatively by negating $absent$.
$$close\_library^{*2}(I, O) \leftarrow weekend(I, O).$$
$$close\_library^{*2}(I, O) \leftarrow not\_absent(I, O).$$

where $weekend(I, O)$ is similarly defined as in the above rule of $weekend^*(I, O)$, and $not\_absent(I, O)$ is defined by the dual transformation for $absent$.

Consider now query ?- $not\_close\_library([\ ], O)$. Provided the dual rules for $close\_library$ and $absent$, we obtain a single solution $[weekend^*, strike^*]$. Note that $insert\_abducible$ rules out $[weekend^*, weekend]$ as a solution, due to its inconsistency.

Abduction with Integrity Constraints

Example 5. Recall Example 1 plus new information: librarians may also be free from working in case the library is being renovated, expressed as:

$$absent \leftarrow renov.$$

Query ?- $close\_library([\,],T)$ now returns an additional solution, viz. $[weekend^*,$ $renov]$. Let us further suppose that the municipal authority permits any renovations to take place only on weekends, expressed as an IC below:

$$false \leftarrow weekend^*, renov.$$

The queries should now respect the IC, too. That is, it should be conjoined with not_false to ensure that all integrity constraints are satisfied:

$$?-\ close\_library([\,],T), not\_false(T,O).$$

where the dual transformation provides the definition for not_false. Note that the abductive solution for $close\_library$ is further constrained by passing it to the subsequent subgoal not_false for confirmation, via the intermediate context T. Using the TABDUAL approach, we can easily check that $[weekend^*, renov]$ is now ruled out from the solutions.

TABDUAL also takes care of abduction over non-ground programs and programs with loops. Abduction over programs with loops sometimes occur in real applications, e.g., in psychiatric diagnosis, cf. [18] that deals with non-stratified negation: "The patient has an Adjustment Disorder if he does not have Alzheimer's Dementia, and has Alzheimer's Dementia if the patient does not have an Adjustment Disorder", which corresponds to rules $adjustment\_disorder \leftarrow not\ alzheimer\_dementia$ and $alzheimer\_dementia \leftarrow not\ adjustment\_disorder$. TABDUAL also benefits from other XSB's features to foster its practicality, e.g., constructing dual rules by need only. The reader is referred to [54] for details of implementation aspects.

3.2 Combining LP Abduction and Updating

In addition to abduction, an agent may learn new knowledge from the external world or update itself internally on its own decision in order to pursue its present goal. It is therefore natural to accommodate abduction systems with knowledge updating. In Section 3.2.1 we summarize an implemented approach for LP updating, called EVOLP/R, that employs incremental tabling [53], an advanced tabling feature of XSB Prolog that lends itself to dynamic environments and evolving systems. Having the two tabling-based approaches for different purposes, viz., TABDUAL and EVOLP/R, we combine them in a unified approach, as discussed in Section 3.2.2.

3.2.1 EVOLP/R for LP Updating

EVOLP/R is theoretically underpinned by Evolving Logic Programs (EVOLP) [2]; the latter is itself based on Dynamic Logic Program [3]. Though its syntax and semantics are based on EVOLP, they differ in several respects, which characterize EVOLP/R, as per below. The reader is referred to [55] for a more detailed theoretical basis of EVOLP/R.

Restriction to Fluent Updates While EVOLP allows full-blown rule updates, EVOLP/R restricts updates to state-dependent literals only (commonly known as *fluents*), i.e., literals whose truth value may change from one state to another. Syntactically, every fluent F is accompanied by its fluent complement $\sim F$. Program updates are enacted by asserting F, whereas its retraction is realized by asserting its complement $\sim F$ at a later state. It thus admits non-monotonicity of a fluent, as the latter update supervenes the former.

Though updates in EVOLP/R are restricted to fluents only, it nevertheless still permits rule updates by introducing a rule name fluent that uniquely identifies the rule for which it is introduced. Such a rule name fluent is placed in the body of a rule to turn the rule on and off, cf. [48]; this being achieved by asserting or retracting that specific fluent. For instance, while EVOLP allows asserting rule *absent ← renov* in Example 5, EVOLP/R does this by introducing a unique rule name fluent, say $rule_{(absent\leftarrow renov)}$, in the body of that rule (this is internally done by EVOLP/R via a program transformation):

$$absent \leftarrow rule_{(absent\leftarrow renov)}, renov. \qquad (5)$$

and asserting the rule name fluent $rule_{(absent\leftarrow renov)}$ at state T, making it true at T and thus turning the rule on. On the other hand, asserting $\sim rule_{(absent\leftarrow renov)}$ at a subsequent state $T' > T$, making the rule name fluent false later at T', and hence turning the rule off from then on.

Incremental Tabling of Fluents The first approach of EVOLP/R [56] preliminarily exploits the combination of two tabling features in XSB Prolog: (1) *incremental tabling* [53], which ensures the consistency of answers in tables with all dynamic facts and rules upon which the tables depend; and (2) *answer subsumption* [63], which allows tables to retain only answers that subsume others wrt. some order relation.

Incremental tabling of fluents is employed in EVOLP/R to automatically maintain the consistency of program states due to assertion and retraction of fluent literals, whether obtained as updated facts (wrt. extensional fluent) or concluded by rules (wrt. to intensional fluent). On the other hand, answer subsumption of fluent literals allows to address the frame problem, by automatically keeping track of only their latest assertion or retraction wrt. the state of a given query. Tabling both a fluent and its complement provides a convenient formulation of counterfactual reasoning (cf. Section 3.5 of [47]).

The combined use of incremental tabling and answer subsumption is realized in the incrementally tabled predicate $fluent(F, Ht_F, Qt)$ for fluent literal F, where Ht_F and Qt are the states (timestamps) when it holds true (*holds-time*) and when it is queried (*query-time*), resp. Invoking $fluent(F, Ht_F, Qt)$ thus, either looks for an entry in its table, if one exists; otherwise, it invokes dynamic definitions of fluent F, and returns the latest time Ht_F fluent F is true wrt. a given query-time Qt. In order to return only the latest time Ht_F when F is true (wrt. Qt), $fluent/3$ is tabled using answer subsumption on its second argument. Predicate $fluent/3$ is then employed in EVOLP/R system predicate $holds(F, Qt)$ to query whether F holds true at query-time Qt. It does so by looking for the entry for F in the table of $fluent(F, Ht_F, Qt)$, obtaining the latest time $Ht_F \leq Qt$, and guarantees that F is not supervened by its complement $\sim F$, i.e., $Ht_{\sim F} \leq Ht_F$, where $Ht_{\sim F}$ is obtained by invoking $fluent(\sim F, Ht_{\sim F}, Qt)$.

While answer subsumption is shown useful in this approach to avoid recursing through the frame axiom by allowing direct access to the latest time when a fluent is true, it requires $fluent/3$ to have query time Qt as its argument. Consequently, it may hinder the reusing of tabled answers of $fluent/3$ by similar goals which differ only in their query-time. Ideally, the state of a fluent literal in time depends solely on the changes made to the world, and not on whether that world is being queried.

The above issue is addressed in the second approach [55], where the use of incremental tabling in EVOLP/R is fostered further, while leaving out the problematic use of answer subsumption. The main idea, not captured in the first approach, is the perspective that knowledge updates (either self or world wrought changes) occur whether or not they are queried: the former take place independently of the latter, i.e., when a fluent is true at Ht, its truth lingers on independently of Qt. Consequently, from the standpoint of the tabled $fluent$ predicate definition, Qt no longer becomes its argument: we now have incremental tabled predicate $fluent(F, Ht_F)$. Invoking $fluent(F, Ht_F)$ thus amounts to look for an entry in its table, if one exists; otherwise, it invokes dynamic definitions of fluent F, and returns a holds-time Ht_F (a state when fluent F is true). Since answer subsumption is left out, more than one instance of holds-time Ht_F of fluent F may be tabled.

The implementation details of EVOLP/R are beyond the scope of this paper, but can be found in [55]. We only illustrate, in Example 6, how the state information, viz., holds-time, figures in a rule transform.

Example 6. Recall rule (5): $absent \leftarrow rule_{(absent \leftarrow renov)}, renov$. For the purpose of this example, assume that *renov* is not an abducible, but instead a fluent. Its rule transform is shown below (abstracted away from implementation details):

$$absent(T) \leftarrow fluent(rule_{(absent \leftarrow renov)}, T_1), \ fluent(renov, T_2), \ latest([T_1, T_2], T).$$

where the reserved predicate *latest* determines a holds-time T of fluent *absent* in the head by which inertial fluent in its body (i.e., fluents $rule_{(absent \leftarrow renov)}$ and *renov*) holds the latest.

The main characteristics of the second approach are summarized below.

- Though more than one instance of holds-time Ht_F of fluent F may be tabled, this approach still permits avoiding top-down recursion or bottom-up iteration through the frame axiom. This is taken care of by the underlying tabling mechanism: look-up a collection of states a fluent literal is true in the table, pick-up the most recent one, and ensure that its complement, with a later state, does not exist in the table. More precisely, querying whether fluent F holds true at query-time Qt amounts to looking for entries for fluent F in the table, by invoking $fluent(F, Ht_F)$, in order to obtain the highest state $Ht_F \leq Qt$, and to guarantee that F is not supervened by its complement $\sim F$, by invoking $fluent(\sim F, Ht_{\sim F})$ and obtaining the highest state $Ht_{\sim F}$ such that $Ht_{\sim F} \leq Ht_F$.

- Propagation of fluent updates is controlled by initially keeping them pending in the database. On the initiative of an actual (top-goal) query, just those updates up to the actual query time are activated, via their incremental assertions, which in turn, automatically trigger system-level incremental bottom-up recomputation of other tabled fluent literals (viz., updating the state information Ht of these fluents). This updating technique thus allows high-level top-down deliberative reasoning about a query to be reconciled with autonomous low-level (via tabling) bottom-up world reactivity to ongoing updates.

- For facilitating the propagation of fluent complements, EVOLP/R borrows the dual transformation from TABDUAL. That is, the transformation derives rules for the dual negation fluent literal $\sim F$ from their corresponding positive rules of fluent F. It thus helps propagate this dual negation fluent literal incrementally, in order to establish whether the fluent or rather its complement is true at some state.

3.2.2 Combining TABDUAL and EVOLP/R

We have presented in the previous sections approaches for LP abduction (TABDUAL) and updating (EVOLP/R). Both approaches enjoy the benefit of tabling features, albeit their different usage. We now recap a unified approach to integrate TABDUAL and EVOLP/R for keeping the benefit of tabling in each individual approach.

Our initial attempt, as reported in [58], employs the concept of hypotheses generation from [42], where the notion of expectation is employed to express preconditions for enabling the assumption of an abducible. An abducible A can be assumed, or *considered*, if there is an expectation for it, and no expectation to the contrary:

$$consider(A) \leftarrow expect(A),\ not\ expect\_not(A),\ A.$$

Note that this concept consequently requires every rule with abducibles in its body to be preprocessed, by substituting abducible A with $consider(A)$. The need and the rule definitions for $expect(A)$ and $expect\_not(A)$ depend on the encoded problem.

While this concept generates only abductive explanations relevant for the problem at hand, the approach in [58] limits the benefit of tabled abduction. That is, instead of associating tabled abduction to a particular atom (e.g., $storm_{ab}$ in Example 3) for a subsequent reuse in another context, a single (and generic) tabled predicate $consider_{ab}$ is introduced for tabling only this single abducible A being considered. Though $expect(A)$ and $expect\_not(A)$ may have rules stating conditions for expecting a or otherwise, it is unnatural that such rules would have other abducibles as a condition for abducing A, as $consider/1$ concerns conditions for *one* abducible only, according to its intended semantics and use. That is, a $consider/1$ call will not depend on another $consider/1$.

Such a generic tabling by $consider\_ab$ is indeed unnecessary and not intended by tabled abduction. In this section we remedy the joint tabling approach of [58] in order to uphold the idea and the benefit of tabled abduction. As in [58], the joint tabling approach in combining TABDUAL and EVOLP/R naturally depends on two pieces of information carried from each of these techniques, viz. abductive contexts and the state when a fluent holds true, respectively. They keep the same function in the integration as in their respective individual approach. Example 7 shows how both entries figure in a remedied rule transform.

Example 7. Recall rule $hit\_lightning \leftarrow storm, low\_altitude$ in Example 2. Its rule transform is shown below (abstracted away from its rule name and other implementation details):

$$hit\_lightning_{ab}(E_2, T) \leftarrow low\_altitude([\ ], E_1, T_1), storm(E_1, E_2, T_2),$$
$$latest([T_1, T_2], T). \tag{6}$$

where E_2 is an abductive solution tabled by predicate $hit\_lightning_{ab}$, obtained by relaying the ongoing abductive solution in context E_1 from subgoal $low\_altitude$ to subgoal $storm$ in the body, given the empty input abductive context (i.e., [ ]) of $low\_altitude$. As in Example 6, the reserved predicate $latest$ determines the state T of $hit\_lightning$ from $low\_altitude$ and $storm$ that latest holds.

There are two points to note in the above remedied transformation. First, the abducible $low\_altitude$ in the body is now called explicitly (rather than immediately placed as the input abductive context of $storm$, cf. rule (1) in Example 3). This is to anticipate possible updates on this abducible. Such an abducible update may take place when one wants to commit to a preferred explanation and fix it in the program as a fact, a case that we show in Section 4 to fix an abduced background context of a counterfactual. Having an abducible as an explicit subgoal in the body thus facilitates bottom-up propagation of its updates (induced by incremental tabling in EVOLP/R), due to the apparent dependency between the tabled predicate in the head (e.g., $hit\_lightning_{ab}$), and the abducible goal in the body (e.g., $low\_altitude$). Second, by prioritizing abducible goals in the body (to occur before any other non-abducible goals), TABDUAL's benefit, of reusing tabled abductive solutions from one context to another, is still obtained. In Example 7, calling abducible $low\_altitude$ with the empty input abductive context – solved using its definition similar to rule (4) – provides $storm$ with an actual input abductive context $E_1 = [low\_altitude]$. It thus achieves the same effect as simply having $[low\_altitude]$ as the input context of $storm$ (cf. rule (1) of Example 3). Using a similar transform as in rule (6), the rule transform for $crash_{ab}$ is given below:

$$crash_{ab}(E_1, T) \leftarrow hit\_lightning([\,], E_1, T_1), storm(E_1, E_2, T_2), latest([T_1, T_2], T).$$

The tabled solution in $hit\_lightning_{ab}$ (rule (6)) can be reused via the definition $hit\_lightning$ below (this definition is similar to rule (3) in Section 3.1, just adding the state information T):

$$hit\_lightning(I, O, T) \leftarrow hit\_lightning_{ab}(E, T), produce\_context(O, I, E).$$

We may observe that, like in Example 3 of TABDUAL, the subgoal $storm$ in the above rule of $crash_{ab}$ is called with a different actual input abductive context (viz., $E_1 = [cumulonimbus, low\_altitude]$) from that of the same subgoal in rule (6).

Finally, the different purposes of the dual program transformation, employed both in TABDUAL and EVOLP/R, are now consolidated within this unified approach:

- Following the name convention for dual predicates in TABDUAL, the dual negation $\sim F$ of fluent literal F is now renamed to not_F in the unified approach.

- As in the positive rule transform, the abductive context and state information (viz., holds-time) jointly figure in the parameters of dual predicates. Recall Example 4. Considering *close_library* as a fluent, the dual transformation of the unified approach results in rules below (abstracted away from the two rule name fluents of *close_library* and other implementation details):

$$not\_close\_library(C_0, C_2, Ht) \leftarrow close\_library^{*1}(C_0, C_1, Ht_1),$$
$$close\_library^{*2}(C_1, C_2, Ht_2),$$
$$latest([Ht_1, Ht_2], Ht) \qquad (7)$$

$$close\_library^{*1}(I, O, Ht_{we^*}) \leftarrow weekend^*(I, O, Ht_{we^*}).$$
$$close\_library^{*2}(I, O, Ht_{we}) \leftarrow weekend(I, O, Ht_{we}).$$
$$close\_library^{*2}(I, O, Ht_{na}) \leftarrow not\_absent(I, O, Ht_{na}).$$

From the abduction perspective, it helps to efficiently deal with downwards by-need abduction under negated goals, e.g., abduction via *close_library*^{*i} due to invoking the goal *not_close_library*$([\,], O, T)$. The state when an abducible is abduced, e.g., Ht_{we^*} of *weekend** in the rule of *close_library*^{*1}, is determined in rule (7) by Ht (viz., by the reserved predicate *latest*): it can be a concrete timestamp Ht_2 or the actual query-time Qt (if Ht_2 is not concrete yet, which is the case when Ht_2 itself refers to another abduction state).

From the updating perspective, it helps to incrementally propagate upwards the dual negation complement of a fluent. For instance, updating the program with *weekend** at Ht_{we^*} will incrementally propagate Ht_{we^*} to determine Ht of fluent *not_close_library*, in this case via *close_library*^{*1}.

4 Counterfactuals with LP Abduction and Updating

In this section we propose an approach that makes use of LP abduction and updating for counterfactual reasoning [47]. It is based on Pearl's approach [40] to evaluate counterfactuals, and adapts it to LP. In Pearl's approach, counterfactuals are evaluated based on a probabilistic causal model and a calculus of intervention. Its main idea is to infer background circumstances that are conditional on current evidences, and subsequently to make a minimal required intervention in the current causal model, so as to comply with the antecedent condition of the counterfactual. The modified model serves as the basis for computing the counterfactual consequent's probability.

Our LP-based approach abstains from the use of probability, and concentrates on pure non-probabilistic counterfactual reasoning in LP, by resorting to LP abduction and updating, to determine the logical validity of counterfactuals under

Well-Founded Semantics (WFS). Nevertheless, the approach is also adaptable to other semantics, e.g., Weak Completion Semantics [24] is employed in [44].

In the sequel, the Well-Founded Model of program P is denoted by $WFM(P)$. The logical consequence relation $\models$ is defined such that for formula F, $P \models F$ iff F is true in $WFM(P)$.

4.1 Causation and Intervention in LP

Two important ingredients in Pearl's approach of counterfactuals are causal model and intervention. Given that the inferential arrow in a LP rule representing causal direction, causation can be captured by LP abduction. It expresses a causal source by providing an explanation to a given observation. With respect to Pearl's causal model: (1) An observation O corresponds to Pearl's definition for evidence. That is, O has rules concluding it in program P, and hence does not belong to the set of abducibles A; (2) Pearl's model comprises a set of background variables (also known as exogenous variables), whose values are conditional on case-considered observed evidences and are not causally explained in the model. In terms of LP abduction, they correspond to a set of abducibles $E \subseteq A$ that provide abductive explanations to observation O. Indeed, these abducibles have no preceding causal explanatory mechanism, as they have no rules concluding them in the program.

Whereas abduction permits obtaining explanations to observations, the evaluation of counterfactual "if Pre had been true, then $Conc$ would have been true", following Pearl's approach, requires the so-called intervention. This is achieved by explicitly imposing the desired truth value of Pre, and subsequently checking whether the predicted truth value of $Conc$ consistently follows from this intervention. As described in Pearl's approach, such an intervention establishes a required adjustment, so as to ensure that the counterfactual's antecedent be met. It permits the value of the antecedent to differ from its actual one, whilst maintaining the consistency of the modified model.

LP updating complements LP abduction in establishing an apposite adjustment to the causal model by *hypothetical* updates of causal intervention on the program, affecting defeasible rules. Moreover, LP updating facilitates fixing the initial abduced background context of the counterfactual being evaluated in the program, by updating the program with the preferred explanation to the current observations.

Next, we detail the roles of LP abduction and updating in our procedure for evaluating the validity of counterfactuals.

4.2 Evaluating Counterfactuals in LP

The procedure to evaluate counterfactuals in LP essentially takes the three-step process of Pearl's approach as its reference. That is, each step in the LP approach captures the same idea of its corresponding step in Pearl's. The three-step procedure is detailed as follows.

Let program P encode the modeled situation on which counterfactuals are evaluated. Consider a counterfactual "if Pre had been true, then $Conc$ would have been true", where Pre and $Conc$ are finite conjunctions of literals.

1. **Abduction:** Perform abduction to explain past circumstances in the presence of evidence (i.e., factual observation).[2] More to the point, compute an explanation $E \subseteq A$ to the observation O. The selected explanation is fixed (via updating) as the abduced background context in which the counterfactual is evaluated ("all other things being equal"), to obtain program $P \cup E$.

2. **Action:** For each literal L in conjunction Pre, introduce a pair of reserved meta-predicates $make(B)$ and $make\_not(B)$, where B is the atom in L. These two meta-predicates are introduced for the purpose of establishing causal intervention: they are used to express hypothetical alternative events to be imposed. This step comprises two stages:

    (a) *Transformation:*
        - Add rule $B \leftarrow make(B)$ to program $P \cup E$.
        - Add *not make_not*(B) to the body of each rule in P whose head is B. If there is no such rule, add rule $B \leftarrow not\ make\_not(B)$ to program $P \cup E$.

       Let $(P \cup E)_\tau$ be the resulting transform.

    (b) *Intervention:* Adjust the resulting transform $(P \cup E)_\tau$ to comply with the antecedent of the counterfactual. The antecedent of the counterfactual is imposed with an intervention on the program, by updating the program with the hypothetical fluents corresponding to the required intervention: update program $(P \cup E)_\tau$ with literal $make(B)$ or $make\_not(B)$, for $L = B$ or $L = not\ B$, resp. Assuming that Pre is consistent, $make(B)$ and $make\_not(B)$ will not be imposed at the same time.
       Let $(P \cup E)_{\tau,\iota}$ be the program obtained after these hypothetical updates of intervention.

---

[2] Factual observations are those explicitly given, and not presupposed from the considered counterfactual. Implicit presuppositions, either those appearing in counterfactual or indicative conditionals, cannot immediately be regarded as observations.

3. **Prediction:** Verify whether the consequent of the counterfactual deductively follows: $(P \cup E)_{\tau,\iota} \models Conc$. Additionally, ensure that all ICs (if any) are also satisfied.

The counterfactual "if Pre had been true, then $Conc$ would have been true" is thus *valid* given an observation O iff O is explained by $E \subseteq A$, $(P \cup E)_{\tau,\iota} \models Conc$, and all ICs (if any) are satisfied in $WFM((P \cup E)_{\tau,\iota})$.

We illustrate in Example 8 how this three-step procedure is actually realized using a combination of LP abduction and updating to evaluate the validity of counterfactuals. The reader is referred to [47] for its formal procedure and implementation details. The procedure has been implemented in a prototype, QUALM, on top of our unified approach of LP abduction and updating (cf. Section 3.2.2), in XSB Prolog.

Example 8. Recall the example in Section 1: *Lightning hits a forest and a devastating forest fire breaks out. The forest was dry after a long hot summer and many acres were destroyed.* Let us consider more abductive causes for the forest fire: storm (which implies lightning hitting the ground) or barbecue. Note that dry leaves are important for forest fire in both cases. This example is expressed in the program below, where the set of abducibles is $A = \{storm, barbecue, storm^*, barbecue^*\}$:

$$fire \leftarrow barbecue, dry\_leaves. \qquad fire \leftarrow barbecue^*, lightning, dry\_leaves.$$
$$lightning \leftarrow storm. \qquad dry\_leaves.$$

The explicit use of $barbecue^*$ in the second rule of $fire$ is intended so as to have mutual exclusive explanations.

Consider counterfactual "if only there had not been lightning, then the forest fire would not have occurred", where the factual observations is $O = \{lightning, fire\}$. Note that the observations assure us that both the antecedent and the consequent literals of the counterfactual were factually false, ensuring that the latter is relevant.

1. **Abduction**: Abduce consistent explanations $E \subseteq A$ to the above factual observations O, viz., $E_1 = \{storm, barbecue^*\}$ and $E_2 = \{storm, barbecue\}$. Say E_1 is preferred for consideration. The abduced background context for the counterfactual is fixed by updating the program with E_1, obtaining $P \cup E_1$.

2. **Action**: The transformation results in program $(P \cup E_1)_\tau$:

$$fire \leftarrow barbecue, dry\_leaves. \qquad fire \leftarrow barbecue^*, lightning, dry\_leaves.$$
$$lightning \leftarrow storm, not\ make\_not(lightning). \qquad dry\_leaves.$$
$$lightning \leftarrow make(lightning).$$

Program $(P \cup E_1)_\tau$ is updated with $make\_not(lightning)$ as the required causal intervention, viz., "if there had not been lightning".

3. **Prediction**: Verify if the conclusion "the forest fire would not have occurred" holds in WFS. Indeed, $(P \cup E_1)_{\tau,\iota} \models not\ fire$. That is, $not\ fire$ holds with respect to the intervened modified program for explanation $E_1 = \{storm, barbecue^*\}$ and the intervention $make\_not(lightning)$. Thus, the counterfactual is valid.

Example 9. Continuing Example 8, if E_2 is instead preferred to update P, the counterfactual is no longer valid. In this case, $(P \cup E_1)_\tau = (P \cup E_2)_\tau$, and the required causal intervention is also the same: $make\_not(lightning)$. But we now have $(P \cup E_2)_{\tau,\iota} \not\models not\ f$. Indeed, this conforms with our understanding that the forest fire would still have occurred but due to an alternative cause, viz., a barbecue.

Skeptical and credulous counterfactual evaluations could ergo be defined, i.e., by evaluating the preferred counterfactual for each abduced background context. Given that step 2 can be accomplished by a one-time transformation, such skeptical and credulous counterfactual evaluations require only executing step 3 for each background context fixed in step 1.

5 Applications to Computational Morality

Counterfactual theories are very suggestive of a conceptual relationship to a form of debugging, namely in view of correcting moral blame, since people ascribe abnormal antecedents an increased causal power, and are also more likely to generate counterfactuals concerning abnormal antecedents. Two distinct processes can be identified when people engage in counterfactual thinking. For one, its frequent spontaneous triggers encompass bad outcomes and "close calls" (some harm that was close to happening). Second, such thinking comprises a process of finding antecedents which, if mutated, would prevent the bad outcome from arising. When people employ counterfactual thinking, they are especially prone to change abnormal antecedents, as opposed to normal ones. Following a bad outcome, people are likely to conceive of the counterfactual "if only [some abnormal thing] had not occurred, then the outcome would not have happened". See [51] for a review. Such a conception of counterfactuals in morality is discussed below by examining the issues of moral permissibility (Section 5.1) and its justification (Section 5.2).

5.1 Moral Permissibility

In this section we look into the application of counterfactuals for examining viewpoints on moral permissibility, exemplified by classic moral dilemmas from the literature on the Doctrines of Double Effect (DDE) [37] and of Triple Effect (DTE)

[29]. Examining moral permissibility with counterfactuals according to these two well-known doctrines does not require formalizing them, but instead detailing, according to our approach, their materialization in concrete moral dilemmas. Classic dilemmas regarding these doctrines are presented, so the results of counterfactual evaluation are readily comparable to those intended in the literature. Moreover, our applications focus on showing the process of moral reasoning through counterfactuals, rather than representing formally the notions of obligation, prohibition, and permissibility in a formal logic (say, as deontic logic operators). Such distinct research direction has been discussed elsewhere, e.g., [49, 7].

DDE is often invoked to explain the permissibility of an action that causes a harm by distinguishing whether this harm is a mere *side-effect* of bringing about a good result, or rather a *means* to bringing about the same good end [37]. In [22], DDE has been utilized to explain the consistency of judgments, shared by subjects from demographically diverse populations, on a series of moral dilemmas.

Counterfactuals may provide a general way to examine DDE in dilemmas, e.g., the classic trolley problem [16], by distinguishing between a *cause* and a *side-effect* as a result of performing an action to achieve a goal. This distinction between causes and side-effects may explain the permissibility of an action in accordance with DDE. That is, *if some morally wrong effect E happens to be a cause for a goal G that one wants to achieve by performing an action A, and not a mere side-effect of A, then performing A is impermissible*. This is expressed by the counterfactual form below, in a setting where action A is performed to achieve goal G:

If not E had been true, then not G would have been true.

The evaluation of this counterfactual form identifies permissibility of action A from its effect E, by identifying whether the latter is a necessary cause for goal G or a mere side-effect of action A. That is, if the counterfactual proves valid, then E is instrumental as a cause of G, and not a mere side-effect of action A. Since E is morally wrong, achieving G that way, by means of A, is impermissible; otherwise, not. Note that the evaluation of counterfactuals in this application is considered from the perspective of agents who perform the action, rather than from others' (e.g., observers). Moreover, our emphasis on causation in this application focuses on agents' deliberate actions, rather than on causation and counterfactuals in general. See [9, 40, 23] for a more general and broad discussion on causation and counterfactuals.

Examples 10 and 11 apply this counterfactual form in two off-the-shelf military cases from [61]: terror bombing vs. tactical bombing. The former refers to bombing a civilian target during a war, thus killing civilians, in order to terrorize the enemy, and thereby get them to end the war. The latter case is attributed to bombing a military target, which will effectively end the war, but with the foreseen consequence of killing the same number of civilians nearby. According to DDE, terror bombing fails permissibility due to a deliberate element of killing civilians to achieve the goal of ending the war, whereas tactical bombing is accepted as permissible.

Example 10. Terror bombing is expressed by the program below, where abducibles are $A_{teb} = \{terror\_bombing, terror\_bombing^*\}$:

$$end\_war \leftarrow terrorize\_enemy. \qquad terrorize\_enemy \leftarrow kill\_civilian.$$
$$kill\_civilian \leftarrow bomb\_civilian. \qquad bomb\_civilian \leftarrow terror\_bombing.$$

We consider end_war as the goal and the counterfactual "if civilians had not been killed, then the war would not have ended", provided the factual observation $O = \{kill\_civilian, end\_war\}$. It has a single explanation $E_{teb} = \{terror\_bombing\}$. Given the transform $kill\_civilian \leftarrow bomb\_civilian, not\ make\_not(kill\_civilian)$ and the intervention $make\_not(kill\_civilian)$, the counterfactual is valid, because $not\ end\_war$ holds in WFS with respect to the intervened modified program and the background context E_{teb}, i.e., $(P \cup E_{teb})_{\tau, \iota} \models not\ end\_war$. That means the morally wrong $kill\_civilian$ is instrumental in achieving the goal end_war: it is a cause for end_war by performing $terror\_bombing$ and not merely its side-effect. Hence $terror\_bombing$ is DDE morally impermissible.

Example 11. Tactical bombing with the same goal end_war is modeled by the program below, where $A_{tab} = \{tactical\_bombing, tactical\_bombing^*\}$:

$$end\_war \leftarrow bomb\_military. \qquad bomb\_military \leftarrow tactical\_bombing.$$
$$kill\_civilian \leftarrow tactical\_bombing.$$

The counterfactual is the same and now we have $E_{tab} = \{tactical\_bombing\}$ as the only explanation to the same observation $O = \{kill\_civilian, end\_war\}$. By imposing the intervention $make\_not(kill\_civilian)$ on the rule transform $kill\_civilian \leftarrow tactical\_bombing, not\ make\_not(kill\_civilian)$, one can verify that the counterfactual is not valid, because end_war holds in WFS with respect to the intervened modified program and the background context E_{tab}. Therefore, the morally wrong $kill\_civilian$ is just a side-effect in achieving the goal end_war. Hence $tactical\_bombing$ is DDE morally permissible.

This application of counterfactuals can be extended to distinguish moral permissibility according to DDE vs. DTE. DTE [29] refines DDE particularly on the notion about harming someone as an intended means. That is, DTE distinguishes further between doing an action *in order* that an effect occurs and doing it *because* that effect will occur. The latter is a new category of action, which is not accounted for in DDE. Though DTE also classifies the former as impermissible, it is more tolerant to the latter (the third effect), i.e., it treats as permissible those actions performed just *because* instrumental harm will occur. Kamm [29] proposed DTE to accommodate a variant of the trolley problem, viz., the *Loop Case* [65]: *A trolley is headed toward five people walking on the track, and they will not be able to get off the track in time. The trolley can be redirected onto a side track, which loops back towards the five. A fat man sits on this looping side track, whose body will by itself stop the trolley. Is it morally permissible to divert the trolley to the looping side track, thereby hitting the man and killing him, but saving the five?* This case strikes most moral philosophers that diverting the trolley is permissible [39]. Referring to a psychology study [22], 56% of its respondents judged that diverting the trolley in this case is also permissible. To this end, DTE may provide the justification, that it is permissible because it will hit the man, and not in order to intentionally hit him [29]. Nonetheless, DDE views diverting the trolley in the Loop case as impermissible.

We use counterfactuals to capture the distinct views of DDE and DTE in the Loop case.

Example 12. We model the Loop case with the program below, where *save, divert, hit, tst, mst* stand for *save the five, divert the trolley, man hit by the trolley, train on the side track* and *man on the side track*, respectively, with *save* as the goal and abducibles $A_{loop} = \{divert, divert^*\}$:

$$save \leftarrow hit. \qquad hit \leftarrow tst, mst. \qquad tst \leftarrow divert. \qquad mst.$$

DDE views diverting the trolley impermissible, because this action redirects the trolley onto the side track, thereby hitting the man. Consequently, it prevents the trolley from hitting the five. To come up with the impermissibility of this action, it is required to show the validity of the counterfactual "if the man had *not* been hit by the trolley, the five people would *not* have been saved". Given the factual observation $O = \{hit, save\}$, its only explanation is $E_{loop} = \{divert\}$. Note that rule $hit \leftarrow tst, mst$ transforms into $hit \leftarrow tst, mst, not\ make\_not(hit)$, and the required intervention is $make\_not(hit)$. The counterfactual is therefore valid, because *not save* holds in WFS of the intervened modified program, given the background context E_{loop}. This means *hit*, as a consequence of action *divert*, is instrumental as a cause of goal *save*. Therefore, *divert* is DDE morally impermissible.

DTE considers diverting the trolley as permissible, since the man is already on the side track, without any deliberate action performed in order to place him there. In Example 12, we have the fact *mst* ready, without abducing any ancillary action. The validity of the counterfactual "if the man had not been on the side track, then he would not have been hit by the trolley", which can easily be verified, ensures that the unfortunate event of the man being hit by the trolley is indeed the consequence of the man being on the side track. The lack of deliberate action (exemplified here by pushing the man – *push* for short) in order to place him on the side track, and whether the absence of this action still causes the unfortunate event (the third effect) is captured by the counterfactual "if the man had *not* been pushed, then he would not have been hit by the trolley". This counterfactual is not valid, because the factual observation $O = \{push, hit\}$ has no explanation $E \subseteq A_{loop}$, i.e., $push \notin A_{loop}$, and no fact $push$ exists either. This means that even without this hypothetical but unexplained deliberate action of pushing, the man would still have been hit by the trolley (just because he is already on the side track). Though hit is a consequence of $divert$ and instrumental in achieving $save$, no deliberate action is required to cause mst, in order for hit to occur. Hence $divert$ is DTE morally permissible.

Example 13. Consider a variant of the Loop case, viz., the *Loop-Push Case* (cf. Extra Push Case in [29]). Differently from the Loop case, now the looping side track is initially empty, and besides the diverting action, an ancillary action of pushing a fat man in order to place him on the side track is additionally performed. This case is modeled by the program below, where $A_{lp} = \{divert, push, divert^*, push^*\}$:

$$save \leftarrow hit. \quad hit \leftarrow tst, mst. \quad tst \leftarrow divert. \quad mst \leftarrow push.$$

Recall the counterfactuals in the discussion of DDE and DTE of the Loop case:

- "If the man had not been hit by the trolley, the five people would not have been saved." The observation $O = \{hit, save\}$ provides an extended explanation $E_{lp_1} = \{divert, push\}$, i.e., the pushing action needs to be abduced for having the man on the side track, so the trolley can be stopped by hitting him. The same intervention $make\_not(hit)$ is applied to the same transform, resulting in a valid counterfactual, because $not\ save$ holds in WFS of the intervened modified program with the background context E_{lp_1}.

- "If the man had not been pushed, then he would not have been hit by the trolley." The factual observation is $O = \{push, hit\}$, explained by $E_{lp_2} = \{divert, push\}$. As there is no rule for $push$, rule $push \leftarrow not\ make\_not(push)$ is introduced, and the intervention $make\_not(push)$ renders $not\ hit$ to hold in WFS of the intervened modified program, given E_{p_2} as its background context. Thus, whereas this counterfactual is not valid in DTE of the Loop case, it is valid in the Loop-Push case.

From the validity of these two counterfactuals it can be inferred that, given the diverting action, the ancillary action of pushing the man onto the side track causes him to be hit by the trolley, which in turn causes the five to be saved. In the Loop-Push, DTE agrees with DDE that such a deliberate action (pushing) performed in order to bring about harm (the man hit by the trolley), even for the purpose of a good or greater end (to save the five), is likewise impermissible.

5.2 Moral Justification

Counterfactuals may as well be suitable to address moral justification, via 'compound counterfactuals': *Had I known what I know today, then if I were to have done otherwise, something preferred would have followed.* Such counterfactuals, typically imagining alternatives with worse effect – the so-called *downward counterfactuals* [35], may provide moral justification for what was done due to a lack in the current knowledge. This is accomplished by evaluating what would have followed if the intent would have been otherwise, other things (including present knowledge) being equal. It may justify that what would have followed is no morally better than the actual ensued consequence.

Example 14. Consider a scenario developed from the Loop case of the trolley problem, which happens on a particularly foggy day. Due to the low visibility, the agent saw only part of the looping side track, so the side track appeared to the agent rather as a straight non-looping one. The agent was faced with a situation whether it was permissible for him to divert the trolley. The knowledge base of the agent with respect to this scenario is shown in a simplified program below. Note that $divert(X)$ is an abducible atom.

$$
\begin{aligned}
run\_sidetrack(X) &\leftarrow divert(X). \\
hit(X,Y) &\leftarrow run\_sidetrack(X), on\_sidetrack(Y). \\
save\_from(X) &\leftarrow sidetrack(straight), run\_sidetrack(X). \\
save\_from(X) &\leftarrow sidetrack(loop), hit(X,Y), heavy\_enough(Y). \\
sidetrack(straight) &\leftarrow foggy. \\
sidetrack(loop) &\leftarrow not\ foggy. \\
foggy. \quad &\quad on\_sidetrack(man). \quad heavy\_enough(man).
\end{aligned}
$$

Taking $save\_from(trolley)$ as the goal, the agent performed counterfactual reasoning "if the man had not been hit by the trolley, the five people would not have been saved". Given the abduced background context $divert(trolley)$, one can verify that the counterfactual is not valid. That is, the man being hit by the trolley is just a side-effect of achieving the goal, and thus $divert(trolley)$ is morally permissible according to DDE. Indeed, this case resembles the original trolley problem (also commonly known as the Bystander case).

At some later time point, the fog has subsided, and by then it was clear to the agent that the sidetrack was looping to the main track. This is achieved by updating the program with *not foggy*, rendering $sidetrack(loop)$ true. There are two standpoints on how the agent can justify his action $divert(trolley)$. For one, it can employ the aforementioned form of compound counterfactual "Had I known that the sidetrack is looping, then if I had not diverted the trolley, the five would have been saved" as a form of self-justification. Given the present knowledge that the side track is looping, the inner counterfactual is not valid, meaning that to save the five people, diverting the trolley (with the consequence of the man being hit) is required. Moreover, the counterfactual "if the man had not been hit by the trolley, the five people would not have been saved", in the abduced context $divert(trolley)$, is valid, meaning that this action is DDE impermissible (cf. Example 12). Therefore, the agent can justify that what would have followed (given its present knowledge, i.e., $sidetrack(loop)$) is no morally better than the actual one, when there was lack of that knowledge: its decision $divert(trolley)$ at that time was instead DDE permissible.

QUALM can evaluate such compound counterfactuals, thanks to its implemented incremental tabling of fluents (Section 3.2.1). Because fluents and their state information are tabled, events in the past subjected to hypothetical updates of intervention can readily be accessed (in contrast to a destructive database approach, cf. [31]). Indeed, these hypothetical updates take place without requiring any undoing of other fluent updates, from the state those past events occurred in up to the current one; more recent updates are kept in tables and readily provide the current knowledge. The discussion of this technique can specifically be found in Section 3.5 of [47].

A different standpoint from where to justify its action is by resorting to Scanlon's *contractualism* [60]. Scanlon argues that moral permissibility can be addressed through the so-called *deliberative* employment of moral judgments. That is, the question of the permissibility of actions is answered by identifying the justified but defeasible argumentative considerations, and their exceptions. It is based on a view that moral dilemmas typically share the same structure: (1) They concern general principles that in some cases admit exceptions; (2) They raise questions about when those exceptions apply. With this structure, an action is determined impermissible through deliberative employment when there is no countervailing consideration that would justify an exception to the applied general principle. In this vein, for the example we are currently discussing, the DTE serves as the exception to justify the permissibility of the action $divert(trolley)$ when the side track was known to be looping, as shown through counterfactual reasoning in Example 12.

We extend now Example 14 to further illustrate how moral permissibility of actions is justified through defeasible argumentative considerations according to Scanlon's contractualism.

Example 15. As the trolley approached, the agent realized that the man was not heavy enough to stop it. This information is acknowledged by the agent through updating its knowledge base with *not heavy_enough(man)*. But there was a heavy cart on the bridge over the looping sidetrack that the agent could push to place it on the side track, and thereby stop the trolley. This scenario is expressed with rules below ($push(X)$ is an abducible atom), in addition to the program of Example 14:

$$on\_sidetrack(X) \leftarrow on\_bridge(X), push(X). \qquad (8)$$

$$on\_sidetrack(Y) \leftarrow push(X), inside(Y, X). \qquad (9)$$

$$on\_bridge(cart). \qquad heavy\_enough(cart).$$

The second rule of $on\_sidetrack$ is an extra knowledge of the agent, that if an object Y is inside the pushed object X, then Y will be on the side track, too.

The goal $save\_from(trolley)$ now succeeds with $[divert(trolley), push(cart)]$ as its abductive solution. But the agent subsequently learned that a fat man, who was heavy enough, was inside the cart: the agent updates its knowledge base with $inside(fat\_man, cart)$ and $heavy\_enough(fat\_man)$. As a consequence, this man was also on the side track and hit by the trolley, which can be verified by query ?- $hit(trolley, fat\_man)$.

In this scenario, a deliberate action of pushing was involved that consequently placed the fat man on the side track (as verified by ?- $on\_sidetrack(fat\_man)$) and the man being hit by the trolley is instrumental to save the five people from the track (as verified by the counterfactual "if the fat man had not been hit by the trolley, the five people would not have been saved"). Nevertheless, the agent may justify the permissibility of its action by arguing that its action is admitted by DTE. In this case, the fat man being hit by the trolley is just a side-effect of the agent's action $push(cart)$ in order to save the five people. Indeed, this justification can be shown through reasoning on the counterfactual "if the cart had not been pushed, then the fat man would not have been hit by the trolley", which is valid given the abduced background context $push(cart)$. Furthermore, the observation $hit(trolley, fat\_man)$ cannot be explained by $push(fat\_man)$ given the absence of the fact $on\_bridge(fat\_man)$, i.e., the hypothetical action $push(fat\_man)$ is not the causal source for the fat man being hit by the trolley.

All moral examples presented in this paper have been successfully tested in QUALM. Examples 14 and 15 together actually form one single program, where QUALM features (derived from its constituents, TABDUAL and EVOLP/R) are exercised. For instance, in Example 14, after updating the program with $not\ foggy$, re-invoking the goal $save\_from(trolley)$ reuses the abductive solution $divert(trolley)$ tabled from the previous invocation of $run\_sidetrack(trolley)$. Moreover, this tabled solution is involved in (as the context of) the deliberative reasoning for the goal $on\_sidetrack(man)$ when $hit(trolley, man)$ is called. It thus provides a computational model of collaborative interaction between deliberative and reactive reasoning in a form of the dual-process model [10, 33], when viewing tabling as a form of low-level reactive behavior. Another feature used, from EVOLP/R, is that rules (8) and (9) are first switched off in Example 14, via the rule name fluent mechanism (cf. Section 3.2.1), as they are not applicable in the considered scenario. Only later, in Example 15, are they switched on again to allow abducing additional action $push(cart)$.

6 Summary and Concluding Remarks

Abduction has recently been on the back burner in LP. One of our research contributions is to address challenges in engineering LP abduction systems, to revive abduction in LP. We present an implemented tabled abduction technique, TABD-UAL, to benefit from LP tabling mechanisms in contextual abduction, so that priorly obtained (and tabled) abductive solutions can be reused from one abductive context to another.

Aiming at facilitating the interplay between LP abduction and other LP non-monotonic reasoning, we have also combined this tabled abduction technique with our own-developed LP updating EVOLP/R; the latter employs an incremental tabling feature of XSB Prolog, and permits a reconciliation of high-level top-down deliberative reasoning about a goal, with autonomous low-level bottom-up world reactivity to ongoing updates.

The other contribution is that of applying our unified approach of LP abduction and LP updating to formulate counterfactuals reasoning based on Pearl's structural theory, but omitting formal probability, given the lack of pure non-probabilistic counterfactual reasoning in LP. In our LP-based counterfactual approach, abduction specifically plays an important role, to explain past circumstances in the presence of evidence (factual observation), and fix (via LP updating) considered explanations as the abduced background context in which the counterfactual is to be evaluated. Our implemented approach and its prototype QUALM is based on the Well-Founded Semantics, but adaptable to other semantics as well, e.g., Weak Completion Semantics [24] is employed in [44], where manual updating is used.

We also apply counterfactuals to model morality issues. Counterfactuals are employed to examine moral reasoning about permissibility by resorting to our LP approach, to distinguish between causes and side-effects as a result of agents' actions to achieve a goal. Counterfactuals are also used to support justifying moral permissibility of agents' actions, either by a compound counterfactual formulation or by arguing an appropriate moral principle as an applicable exception, following

Scanlon's contractualism.

Side-effects in abduction have been investigated in [42, 43] through the concept of inspection points; the latter are construed in a procedure by 'meta-abducing' a specific abducible *abduced(A)* whose function is only that of checking that its corresponding abducible *A* is indeed already adopted elsewhere. Therefore, the consequence of the action that triggers this 'meta-abducing' is merely a side-effect. Indeed, inspection points may be employed to distinguish a cause from a mere side-effect, and thus may provide an alternative or supplement to counterfactuals employed for the above same purpose of examining moral permissibility.

The moral examples in the present paper do not exploit the use of ICs. The use of ICs within abduction, but without counterfactuals, for morality has been explored in our related work [45]. They are particularly useful to rule out impermissible actions. One of the difficulties in using an IC to express impermissibility is that it requires the representation to be crafted in sufficient detail in order for the IC to be applicable. While we use counterfactuals to examine permissibility (so we are not bound to have a subtle problem representation), ICs can be used for other purposes, e.g., to choose among mutually exclusive abducibles. On the other hand, ICs should be treated carefully in counterfactuals, because an intervention may render ICs unsatisfiable, and hence their body's support may need to be abductively revised in order to re-impose satisfaction.

While moral examples in this paper are based on those from the literature with either their conceptual or empirical results readily available, a more systematic evaluation of counterfactual reasoning in computational morality may be considered as future work. For instance, an empirical study with human subjects, say in collaboration with cognitive scientists, may be conducted to provide insights on relevant counterfactuals in examining moral permissibility of considered cases, as well as in examining argumentative processes of moral reasoning in justifying permissibility via counterfactuals. The study will then be able to guide the form of counterfactuals to represent and to reason about using the approach fostered in this paper.

References

[1] M. Alberti, F. Chesani, M. Gavanelli, E. Lamma, and P. Torroni. Security protocols verification in abductive logic programming. In *6th Int. Workshop on Engineering Societies in the Agents World (ESAW)*, volume 3963 of *LNCS*. Springer, 2005.

[2] J. J. Alferes, A. Brogi, J. A. Leite, and L. M. Pereira. Evolving logic programs. In *JELIA 2002*, volume 2424 of *LNCS*, pages 50–61. Springer, 2002.

[3] J. J. Alferes, J. A. Leite, L. M. Pereira, H. Przymusinska, and T. Przymusinski. Dynamic updates of non-monotonic knowledge bases. *Journal of Logic Programming*, 45(1-3):43–70, 2000.

[4] J. J. Alferes, L. M. Pereira, and T. Swift. Abduction in well-founded semantics and generalized stable models via tabled dual programs. *Theory and Practice of Logic Programming*, 4(4):383–428, 2004.

[5] J. Balsa, V. Dahl, and J. G. Pereira Lopes. Datalog grammars for abductive syntactic error diagnosis and repair. In *Proc. Natural Language Understanding and Logic Programming Workshop*, 1995.

[6] C. Baral and M. Hunsaker. Using the probabilistic logic programming language P-log for causal and counterfactual reasoning and non-naive conditioning. In *IJCAI 2007*, 2007.

[7] S. Bringsjord, K. Arkoudas, and P. Bello. Toward a general logicist methodology for engineering ethically correct robots. *IEEE Intelligent Systems*, 21(4):38–44, 2006.

[8] R. M. J. Byrne. *The Rational Imagination: How People Create Alternatives to Reality*. MIT Press, 2007.

[9] J. Collins, N. Hall, and L. A. Paul, editors. *Causation and Counterfactuals*. MIT Press, 2004.

[10] F. Cushman, L. Young, and J. D. Greene. Multi-system moral psychology. In J. M. Doris, editor, *The Moral Psychology Handbook*. Oxford University Press, 2010.

[11] M. Denecker and D. de Schreye. SLDNFA: An abductive procedure for normal abductive programs. In *Procs. of the Joint Intl. Conf. and Symp. on Logic Programming*. The MIT Press, 1992.

[12] J. Dix. A classification theory of semantics of normal logic programs: II. weak properties. *Fundamenta Informaticae*, 3(22):257–288, 1995.

[13] T. Eiter, G. Gottlob, and N. Leone. Abduction from logic programs: semantics and complexity. *Theoretical Computer Science*, 189(1-2):129–177, 1997.

[14] K. Epstude and N. J. Roese. The functional theory of counterfactual thinking. *Personality and Social Psychology Review*, 12(2):168–192, 2008.

[15] K. Eshghi. Abductive planning with event calculus. In *Proc. Intl. Conf. on Logic Programming*. The MIT Press, 1988.

[16] P. Foot. The problem of abortion and the doctrine of double effect. *Oxford Review*, 5:5–15, 1967.

[17] T. H. Fung and R. Kowalski. The IFF procedure for abductive logic programming. *Journal of Logic Programming*, 33(2):151–165, 1997.

[18] J. Gartner, T. Swift, A. Tien, C. V. Damásio, and L. M. Pereira. Psychiatric diagnosis from the viewpoint of computational logic. In *Procs. 1st Intl. Conf. on Computational Logic (CL 2000)*, volume 1861 of *LNAI*, pages 1362–1376. Springer, 2000.

[19] M. L. Ginsberg. Counterfactuals. *Artificial Intelligence*, 30(1):35–79, 1986.

[20] J. Y. Halpern and C. Hitchcock. Graded causation and defaults. British Journal for the Philosophy of Science (forthcoming), Available from http://arxiv.org/abs/1309.

1226, 2014.

[21] T. A. Han, A. Saptawijaya, and L. M. Pereira. Moral reasoning under uncertainty. In *LPAR-18*, volume 7180 of *LNCS*, pages 212–227. Springer, 2012.

[22] M. Hauser, F. Cushman, L. Young, R. K. Jin, and J. Mikhail. A dissociation between moral judgments and justifications. *Mind and Language*, 22(1):1–21, 2007.

[23] C. Hoerl, T. McCormack, and S. R. Beck, editors. *Understanding Counterfactuals, Understanding Causation: Issues in Philosophy and Psychology*. Oxford University Press, 2011.

[24] S. Hölldobler and C. D. P. Kencana Ramli. Logic programs under three-valued Łukasiewicz semantics. In *ICLP 2009*, volume 5649 of *LNCS*, pages 464–478. Springer, 2009.

[25] K. Inoue and C. Sakama. A fixpoint characterization of abductive logic programs. *J. of Logic Programming*, 27(2):107–136, 1996.

[26] A. Kakas, R. Kowalski, and F. Toni. Abductive logic programming. *Journal of Logic and Computation*, 2(6):719–770, 1992.

[27] A. C. Kakas and P. Mancarella. Knowledge assimilation and abduction. In *Intl. Workshop on Truth Maintenance*, ECAI'90, 1990.

[28] A. C. Kakas and A. Michael. An abductive-based scheduler for air-crew assignment. *J. of Applied Artificial Intelligence*, 15(1-3):333–360, 2001.

[29] F. M. Kamm. *Intricate Ethics: Rights, Responsibilities, and Permissible Harm*. Oxford U. P., 2006.

[30] R. Kowalski. *Computational Logic and Human Thinking: How to be Artificially Intelligent*. Cambridge U. P., 2011.

[31] R. Kowalski and F. Sadri. Abductive logic programming agents with destructive databases. *Annals of Mathematics and Artificial Intelligence*, 62(1):129–158, 2011.

[32] D. Lewis. *Counterfactuals*. Harvard U. P., 1973.

[33] R. Mallon and S. Nichols. Rules. In J. M. Doris, editor, *The Moral Psychology Handbook*. Oxford University Press, 2010.

[34] D. R. Mandel, D. J. Hilton, and P. Catellani, editors. *The Psychology of Counterfactual Thinking*. Routledge, New York, NY, 2005.

[35] K. D. Markman, I. Gavanski, S. J. Sherman, and M. N. McMullen. The mental simulation of better and worse possible worlds. *Journal of Experimental Social Psychology*, 29:87–109, 1993.

[36] R. McCloy and R. M. J. Byrne. Counterfactual thinking about controllable events. *Memory and Cognition*, 28:1071–1078, 2000.

[37] A. McIntyre. Doctrine of double effect. In E. N. Zalta, editor, *The Stanford Encyclopedia of Philosophy*. Center for the Study of Language and Information, Stanford University, Fall 2011 edition, 2004. http://plato.stanford.edu/archives/fall2011/entries/double-effect/.

[38] S. Migliore, G. Curcio, F. Mancini, and S. F. Cappa. Counterfactual thinking in moral judgment: an experimental study. *Frontiers in Psychology*, 5:451, 2014.

[39] M. Otsuka. Double effect, triple effect and the trolley problem: Squaring the circle in looping cases. *Utilitas*, 20(1):92–110, 2008.

[40] J. Pearl. *Causality: Models, Reasoning and Inference*. Cambridge U. P., 2009.

[41] L. M. Pereira, J. N. Aparício, and J. J. Alferes. Counterfactual reasoning based on revising assumptions. In *ILPS 1991*, pages 566–577. MIT Press, 1991.

[42] L. M. Pereira, P. Dell'Acqua, A. M. Pinto, and G. Lopes. Inspecting and preferring abductive models. In K. Nakamatsu and L. C. Jain, editors, *The Handbook on Reasoning-Based Intelligent Systems*, pages 243–274. World Scientific Publishers, 2013.

[43] L. M. Pereira, E-A. Dietz, and S. Hölldobler. Contextual abductive reasoning with side-effects. *Theory and Practice of Logic Programming*, 14(4-5):633–648, 2014.

[44] L. M. Pereira, E.-A. Dietz, and S. Hölldobler. An abductive counterfactual reasoning approach in logic programming. Available from http://goo.gl/bxOmIZ, 2015.

[45] L. M. Pereira and A. Saptawijaya. Modelling Morality with Prospective Logic. In M. Anderson and S. L. Anderson, editors, *Machine Ethics*, pages 398–421. Cambridge U. P., 2011.

[46] L. M. Pereira and A. Saptawijaya. Bridging two realms of machine ethics. In J. B. White, editor, *Rethinking Machine Ethics in the Age of Ubiquitous Technology*. IGI Global, 2015.

[47] L. M. Pereira and A. Saptawijaya. Counterfactuals in logic programming with applications to agent morality. In R. Urbaniak and G. Payette, editors, *Applied Formal/Mathematical Philosophy, Logic, Argumentation & Reasoning*. Springer, 2016, 2016.

[48] D. L. Poole. A logical framework for default reasoning. *Artificial Intelligence*, 36(1):27–47, 1988.

[49] T. M. Powers. Prospects for a Kantian machine. *IEEE Intelligent Systems*, 21(4):46–51, 2006.

[50] O. Ray, A. Antoniades, A. Kakas, and I. Demetriades. Abductive logic programming in the clinical management of HIV/AIDS. In *Proc. 17th. European Conference on Artificial Intelligence*. IOS Press, 2006.

[51] N. J. Roese. Counterfactual thinking. *Psychological Bulletin*, 121(1):133–148, 1997.

[52] N. J. Roese and J. M. Olson, editors. *What Might Have Been: The Social Psychology of Counterfactual Thinking*. Psychology Press, New York, NY, 2009.

[53] D. Saha. *Incremental Evaluation of Tabled Logic Programs*. PhD thesis, SUNY Stony Brook, 2006.

[54] A. Saptawijaya and L. M. Pereira. TABDUAL: a tabled abduction system for logic programs. *IfCoLog Journal of Logics and their Applications*, 2(1):69–123.

[55] A. Saptawijaya and L. M. Pereira. Incremental tabling for query-driven propagation of logic program updates. In *LPAR-19*, volume 8312 of *LNCS*, pages 694–709. Springer, 2013.

[56] A. Saptawijaya and L. M. Pereira. Program updating by incremental and answer subsumption tabling. In *LPNMR 2013*, volume 8148 of *LNCS*, pages 479–484. Springer, 2013.

[57] A. Saptawijaya and L. M. Pereira. Tabled abduction in logic programs (Technical Communication of ICLP 2013). *Theory and Practice of Logic Programming, Online Supplement*, 13(4-5), 2013. `http://journals.cambridge.org/downloadsup.php?file=`
`/tlp2013008.pdf`.

[58] A. Saptawijaya and L. M. Pereira. Joint tabling of logic program abductions and updates (Technical Communication of ICLP 2014). *Theory and Practice of Logic Programming, Online Supplement*, 14(4-5), 2014. Available from `http://arxiv.org/abs/`
`1405.2058`.

[59] A. Saptawijaya and L. M. Pereira. The potential of logic programming as a computational tool to model morality. In R. Trappl, editor, *A Construction Manual for Robots' Ethical Systems: Requirements, Methods, Implementations*, Cognitive Technologies. Springer, 2015.

[60] T. M. Scanlon. *What We Owe to Each Other*. Harvard University Press, 1998.

[61] T. M. Scanlon. *Moral Dimensions: Permissibility, Meaning, Blame*. Harvard University Press, 2008.

[62] T. Swift. Tabling for non-monotonic programming. *Annals of Mathematics and Artificial Intelligence*, 25(3-4):201–240, 1999.

[63] T. Swift and D. S. Warren. Tabling with answer subsumption: Implementation, applications and performance. In *JELIA 2010*, volume 6341 of *LNCS*, pages 300–312. Springer, 2010.

[64] T. Swift and D. S. Warren. XSB: Extending Prolog with tabled logic programming. *Theory and Practice of Logic Programming*, 12(1-2):157–187, 2012.

[65] J. J. Thomson. The trolley problem. *The Yale Law Journal*, 279:1395–1415, 1985.

[66] A. van Gelder, K. A. Ross, and J. S. Schlipf. The well-founded semantics for general logic programs. *Journal of ACM*, 38(3):620–650, 1991.

[67] J. Vennekens, M. Bruynooghe, and M. Denecker. Embracing events in causal modeling: Interventions and counterfactuals in CP-logic. In *JELIA 2010*, volume 6341 of *LNCS*, pages 313–325. Springer, 2010.

Received January 2015

A Dynamic Approach to Peirce's Interrogative Construal of Abductive Logic

Minghui Ma[*]

Institute for Logic and Intelligence, Southwest University, Chongqing
mmh.thu@gmail.com

Ahti-Veikko Pietarinen[†]

Center for Epistemology and Cognition, Xiamen University, Xiamen
Peirce Research Centre, University of Helsinki, Helsinki
ahti-veikko.pietarinen@helsinki.fi

Abstract

Peirce's novel, post-1903 interrogative construal of abductive logic is studied in terms of a dynamic approach. The information flow in abduction is interpreted by a dynamic mechanism over ranges. An abductive logic is proposed that analyzes conjecture-making in scientific discovery. It explains how the updates on models representing the inquirers' information proceed when they are faced with new surprising facts. The formal semantics for this logic is given by neighborhood models. We apply our abductive logic to an analysis of some examples of scientific discovery.

We thank the two anonymous reviewers of this journal for the abductive feedback that has contributed so much to the improvement of the present paper. We also thank the audience of the 15th Congress of Logic, Methodology and Philosophy of Science held in Helsinki, August 2015, in which a part of the present paper was presented in a special symposium on *Pragmati(ci)st Philosophy of Science: Old and New* organized by Chiara Ambrosio.

[*]This work is supported by China National Fund of Humanities and Social Sciences (grant no. 14ZDB016).

[†]This work is supported by the Academy of Finland (project 1270335) and the Estonian Research Council (project PUT 267) (*Diagrammatic Mind: Logical and Cognitive Aspects of Iconicity*, Principle Investigator A.-V. Pietarinen).

1   Introduction

Peirce discovered *abductive*—which he also variously termed hypothetical, retroductive, adductive or presumptive—reasoning in his critics on logic he was developing in the late 19th-century. Initially, Peirce conceived deductive logic as the logic of mathematics, and the inductive and abductive logic as the logic of science. Later, he took these three types of reasoning to represent different *stages* of scientific inquiry rather than different *types* of reasoning.

According to Peirce, "the art of making explanatory hypotheses is the supreme branch of logic" (Peirce to J. Royce, June 30, 1913).[1] What is it that such an art harbours? In July 16, 1905, Peirce drafted a letter which he intended to send to Victoria Welby: "This kind of reasoning by which theories are formed under the inspiration of the facts I call *Abduction* ...to warrant my use of the word 'abduction' to denote the process of conjecturally setting up a theory suggested by the study of *surprising* facts to explain their being surprising. It is very remarkable that few logicians have ever attempted to define the circumstances under which an explanatory hypothesis is called for".[2]

In this paper, we develop upon Peirce's unpublished remarks concerning his late, post-1903 development of abduction. His novel point is that one can characterize abduction as "reasoning from surprise to inquiry". Compared with the rule of Modus Tollens, he formulated abduction as follows: given a (surprising) fact C, if A implies C, then *it is to be inquired whether A plausibly* holds. The important discovery is that the conclusion is presented in a kind of interrogative mood. But the interrogative mood does not merely mean that a question is raised. In fact, it means that the possible conjecture A becomes the subject of inquiry: the purpose is to determine whether that A is indeed plausible or not. Peirce termed such mood "the investigand mood". Hence abduction can be viewed as the dynamic process toward a plausible conjecture and, ultimately, toward a limited set of the most plausible conjectures.

We reconstruct Peirce's interrogative construal of abductive logic in terms of a dynamic approach, interpreting the information flow in abduction by a dynamic mechanism over epistemic ranges. The first, preliminary analysis is based on propositional logic. Reasoning is a dynamic process from premises to the conclusion. The interrogative mood shows up as an uncertainty in the final range of truth-value configurations. Although this analysis is far too simple to capture the full meaning of

---

[1]From the Papers of Josiah Royce deposited at the Harvard University Archives, published in [28, pp. 765–766].

[2]This letter was never sent and it has not been published before. It is preserved among the original correspondence [24], but it was omitted from the publication of the Peirce–Welby correspondence [13]. A complete transcription of this letter is to appear in [28, pp. 842–845].

the special mood that is required, in drafting plausible conjectures we nevertheless uncover the dynamic perspective on reasoning, a necessary preliminary step towards a more full-blown logic of abduction.

The second step for achieving the analysis of Peirce's interrogative construal of abduction is to introduce a *logic of making conjectures*. Here conjectures are propositions which are obtained by the agent by guessing from a surprising fact. Some of them may become plausible after a dynamic process of inquiring. We introduce a special interpretation to the modal operator $\Box A$ which can mean, for example, that 'A is conjectured', or alternatively, and historically and conceptually perhaps even more accurately, that 'it is hoped that A'.[3]

The models for our logic are neighborhood models. The advantage of using such models is that one can easily express that a proposition is a conjecture in the epistemic or mental range of the current state. If we were to use standard models of modal logic, we should admit principles of logical omniscience. However, the logic of neighbourhood models does not assume logical omniscience, with the sole exception of closure under logical equivalents and, in some cases, closure under valid implications (when monotonicity and non-emptiness is assumed). Second, using neighbourhood models can represent conjectures in an insightful manner. The semantic value of a proposition is supposed to be a set of possible worlds. In neighbourhood models, to be in a neighbourhood of the current world is to be a conjecture (proposition) at the current world. Third, each standard (Hintikka-Kripke) model is in fact equivalent to a neighbourhood model, but the converse does not hold. This means that ne! ighbourhood models are more general than standard structures. In order to interpret the dynamic process of forming a *plausible* conjecture, we introduce an updating mechanism that changes the neighborhood model into one in which we achieve a plausible conjecture.

The third step we take is to introduce dynamic operators into the logic of conjecture. It is here that we achieve the full dynamic logic of abduction. We introduce the dynamic operator on the modal one, $[!F]\Box A$, which means that A preserves its status as a conjecture after the surprising fact F has occurred. In other words, the conjecture, or the hope that one levies on the hypothesis, survives the experience encountering a series of further surprising facts. The dynamic operator is interpreted by updating a neighborhood model by such new facts F. Of course, a scientist or a research team expects to encounter a vast number of such surprising facts in the course of inquiry. It is the hallmark of frontier sciences that one is on the relentless

---

[3]'To hope' may even be slightly weaker than 'to conjecture'. Certainly it is weaker than belief. Peirce did not take abduction to result in anything stronger than conjecturing: "processes of thought capable of producing no conclusion more definite than a conjecture ... I now call Abduction" (MS 293, 1906-7).

lookout for new sources of information to discover new and surprising facts. In looking for warrants for their hypotheses, scientists seldom rest their case on evidence already in their possession.

The final step of our work is to test the dynamic logic of abduction by two examples, one from Dewey and the other from Peirce.

2 Peirce's Abductive Logic in Perspective

In his early works from the 1860s, Peirce used syllogism to prove that there are only three fundamental types of reasoning: deduction, abduction and induction. These correspond to the following three figures of syllogism:

Figure I. Deduction	Figure II. Abduction	Figure III. Induction
S is M	S is M	M is S
M is P	P is M	M is P
S is P	S is P	S is P

One of Peirce's examples of abduction (termed that early time the hypothetic argument) under Figure II was: Light is polarizable. Ether waves are polarizable. Therefore, light is ether waves (W 1:427, 1866). The form of the hypothetic argument is the only one that enables us, Peirce explains, to "see the *why* of things" (W 1:428).

Later in the late 1870s and 1880s [21, 22], he described abduction and induction as inversions of a deductive syllogism in the following sense. We say that the major premise of a syllogism in Figure I is *Rule*, its minor premise *Case*, and its conclusion *Result*. Then abduction is the inference from a result and a rule to a case, and induction is the inference from a case and a result to a rule. Table 1 below presents three examples of these three types of inference.

Later still, towards the turn of the century Peirce arrived at the view that these three types of inference are in fact three stages in scientific inquiry [30]. The first stage is abduction, by which one or more hypotheses or conjectures that may explain some surprising fact is set forth. The second stage is deduction, by which the necessary consequences of the hypothesis are traced. The third stage is induction, which puts the consequences to test and generalizes its conclusion. Every example of scientific inquiry is for Peirce bound to follow the pattern of *abduction–deduction–induction*. It was moreover shown in [30] that even deduction itself breaks into such patterns: abduction, for instance, appears as a 'moment', or the first phase, of deductive reasoning, which is concerned with definition and logical analysis.[4]

---

[4]The reason for what might strike one as a surprising assertion of deduction also involving

Deduction	
Rule:	All the beans in this bag are white.
Case:	These beans are from this bag.
Result:	These beans are white.
Abduction	
Rule:	All the beans in this bag are white.
Result:	These beans are white.
Case:	These beans are from this bag.
Induction	
Case:	These beans are from this bag.
Result:	These beans are white.
Rule:	All the beans in this bag are white.

Table 1: Figures of Syllogism

Of the three methods, Peirce took deduction to be *the most secure* and the least fertile, while abduction is, in contrast, *the most fertile* and the least secure. Since abduction is not secure, it is that mode of reasoning that is best suited to deal with uncertainty and guess, prevalent in everyday scientific practice. But it therefore also requires some compelling reasons for its validity and justification in order for us to regard it as a trustworthy mode of reasoning in the sciences [30].

The central question of Peirce's late formulation of abductive logic is to interpret it as a logic of scientific discovery. This in view, Peirce came to generalize the syllogistic description of abduction. The following logical form has come to be the standard—and even termed the "classical" one in [31]—description:

> The surprising fact, C, is observed.
> But if A were true, C would be a matter of course.
> Hence, there is reason to suspect that A is true.

The truth is that Peirce happened to deliver this schema relatively casually, during the seventh of his 1903 Harvard Lectures (CP 5.189). This widely popular scheme nevertheless became, quite unwarrantably, the form routinely referred to in the subsequent literature. It by no means was his best or the final formulation of what the logical form of abduction ought to look like. It is telling that after 1903, Peirce

---

abduction is that deduction is not only demonstrative but is importantly reliant on *theorematic* types of reasoning. And it is such theorematic reasoning that comes to the fore in the tasks of logical analysis and defining concepts, both of which Peirce took to pertain to deduction in its first phase, followed by demonstration.

hardly ever refers to it. On the contrary, he goes on to propose new logical forms for abduction, more often in fact termed *retroduction* than abduction.

The matter has consequently caused some serious confusions in the literature. The 1903 schema indeed encounters several problems. Any interpretation of abductive reasoning as the logic of scientific discovery which would exclusively be based on the analysis and application of that 1903 schema would be insufficient or misleading at least for the following reasons:

(1) Recognition of what counts as 'surprising' requires a separate explanation. It involves a special kind of scientific intelligence and experience by which to observe, recognize and identify such facts.

(2) The conditional major premiss is not in an indicative but in a subjunctive mood, making it much harder to be analyzed in semantical and logical precision.

(3) The A's and C's are not singular propositions but *conceivable states of things* (the A's) and *complex facts* (the C's). For instance the design of controlled experiments is an example of the tasks that aim at spelling out the possible dependent relationships between A's and C's.

(4) The schema is an idealized sketch that masks much of the complexity involved in actual processes of scientific discovery, including the role of background theories and research agendas, and the connected questions concerning the economy of research. Peirce attached a great deal of importance to such issues of economy. He for instance remarked that "the whole question of what one out of a number of possible hypotheses ought to be entertained becomes purely a question of economy" (CP 6.528).

(5) The schema might indicate, misleadingly, that the essence of abductive reasoning merely deflates into that of the inference to the best explanation (IBE).

As to the last point, the difference in brief is that abduction allows generating more hypotheses than the IBE account does. Abduction is the method for theory building, while IBE compares explanatory values of hypotheses. IBE has no account of what a scientific guess may be, although the latter is basically tantamount to Peirce's conception of abduction. Reasons for keeping the two clearly separate are also found in how abduction copes with scientific uncertainties and ignorance that falls outside of standardized probabilistic methods.

This last, deflationist assimilation might even suggest some a reduction of abduction to Bayesian, confirmational modes of reasoning. That would be an outright wrong direction to look for the essence of abductive reasoning. It would strip abduction of its most important features that enabled Peirce to delineate it as the only "originary" (CP 5.171) mode of reasoning: reasoning that not only is vaguely classified as creative in suggesting new ideas but also and more importantly, originates by coping with fundamental uncertainties in real discovery. A research project that proposes to cope with such uncertainties and structural ignorance in some novel and unexpected ways may stand a chance of increasing the *plausibility*, but not those of probability, credibility, or degrees of beliefs in hypotheses, thus increasing the possibility of progress in not eliminating the lines of research that the probabilistic assessment schemas would have done at the outset.[5]

Before moving on to the main logical form for abduction, we note that Peirce also gave the following description in one of the late unpublished manuscripts:

> In the inquiry, all the possible significant circumstances of the surprising phenomenon are mustered and pondered, until a conjecture furnishes some possible Explanation of it, by which I mean a syllogism exhibiting the surprising fact as necessarily following from the circumstances of its occurrence together with the truth of the conjecture as premises. (MS 843, p. 41, 1908; cf. CP 6.469–470)

The explaining syllogism here is as follows:

> If A were true, C would be observable.
> A is true.
> Therefore, C is observable.

Here the major premise is in a subjunctive mood, and the reasoning is by Modus Ponens on the subjunctive and the minor premise. We take this kind of reasoning quite unsatisfactory, however, as that applies the MP to a subjunctive conditional.

A highly important logical form of abduction is found in the July 16, 1905 unpublished and unsent letter which Peirce drafted to Victoria Welby. In this form, the major premiss is in an *indicative* and the conclusion in an *interrogative* mood. Peirce explains the interrogative mood as follows:

> [The] "interrogative mood" does not mean the mere idle entertainment of an idea. It means that *it will be wise to go to some expense, dependent upon the advantage* that would accrue from knowing that Any/Some S is M, provided that *expense would render it safe to act on that assumption supposing it to be true*. This is the kind of reasoning called reasoning from consequent to antecedent. For it is related to the *Modus Tollens* thus:

---

[5] See [5, 9, 14, 20, 27, 30, 36] for related criticisms on those that have conflated IBE and abduction in the past, and [29] for a criticism of Bayesian approach from the strategic perspective of game theory. Few genuinely productive research projects have *ex ante* priors to work with in the early stages of inquiry.

ModusTollens	Abduction
If A is true, C $^{\text{is not}}_{\text{is}}$ true	If A is true, C $^{\text{is not}}_{\text{is}}$ true
But C $^{\text{is}}_{\text{is not}}$ true	But C $^{\text{is not}}_{\text{is}}$ true
Therefore, A is not true	Therefore, Is A not true?

Instead of "interrogatory", the mood of the conclusion might more accurately be called "investigand", and be expressed as follows:

"It is to be inquired whether A is not true."

The reasoning might be called "Reasoning from Surprise to Inquiry" (Peirce to Welby, July 16, 1905, added emphases).

The term 'investigand' is highly suggestive here. We can relate it to a multiplicity of issues that constitute the initial course of scientific thought, among them the following six: (i) Noticing the surprising phenomenon, (ii) searching for pertinent circumstances, (iii) asking a question, (iv) forming a conjecture, (v) remarking that the conjecture at least appears to explain the surprising phenomenon, (vi) adopting the conjecture as an imaginable, comprehensible and plausible one, (v) paying close attention to economic factors and cost-benefit analysis in regard to the potential testing of the hypothesis, and (vi) a recommendation for a reasonably safe matter of course to act on those assumptions that suppose certain situations that scientific hypotheses propose. All these phases and aspects of inquiry are constitutive of the first, abductive stage of inquiry. Its heart is that of forming the conjecture.[6]

The forming of conjectures has often been described, and occasionally even by Peirce himself, as an act or a flash of *insight*, or having an *instinct* for guessing right. There are reasons to take conjecture-forming to be a genuinely logical process, however, and the present paper is a contribution to those reasons. Glossing abduction as an act or a flash of insight, or an instinctive form of behaviour, both fail to capture the crucial logical characteristics involved in that mode of reasoning, however, as has been pointed out in [3].

In the 1905 letter, Peirce's remarkable move is to take abduction to be a Modus Tollens from the premises "If A is true, C is not true" and "But C is not true" to the interrogative, "Is A not true?". He describes the process illustratively as

---

[6]A reviewer points out that these aspects of inquiry that we take to pertain to the abductive stage of reasoning are somehow similar to, although definitely more detailed than, what was proposed in [34]. The two papers are similar in that they attempt at a dynamic characterization of abduction in the overall framework of epistemic logic. Where the two approaches differ is epistemological: we interpret the essential modality as "conjecturing" or "hoping that", which is even weaker than belief.

"Reasoning from Surprise to Inquiry". In this reasoning, the mood of the conclusion is a peculiar combination of interrogative and imperative moods; a mood which we may not find in natural languages. Unlike in the 1903 schema, the new schema also is a non-committal to the suspectibility of A's truth. Rather, the commitment is to go ahead with the investigation: "it is to be inquired whether A is not true". Peirce termed this the "investigand" mood. This is a clear development of his 1903 view, according to which abduction "commits us to nothing. It merely causes a hypothesis to be set down upon our docket of cases to be tried" (CP 5.603, 1903), which he now realizes as too poor an account. The 1905 schema is equally a development of his suggestion a few years earlier that the abductive conclusion is a 'Maybe A'. This he now also finds too weak and indecisive, even though a 'May-be' it might come just slightly above the pure possibility in its strength. Abductive conclusions must become committals to action, that something is to be done, while suggesting some real pointers towards what is it that one ought to do. What the 1905 abductive schema does, then, is to embed the key qualities of the economy of research into its logical form. Questions of economy are not extra-logical or supererogatory questions; they are part and parcel of the logical form of abductive reasoning.[7]

We note that it is here in the development that takes place in 1905 that abduction, now conceived as a reasoning process that ranges from the preliminary feelings of surprise to the initiation of large-scale scientific endeavours, comes strikingly close to what Hintikka's interrogative model of inquiry has purported to accomplish [14]. The details of those connections are yet to be spelled out.

What is also important in Peirce's draft letter is that the major premise is no longer presented in the subjunctive mood. While subjunctive moods are surely the hallmarks both of scientific precepts as well as Peirce's philosophy of pragmaticism, our task is simplified as the models for such logical reasoning on indicative conditionals are easier to formulate. These 1905 remarks on the logical form of abduction are our guidelines in our study of the logical form of abduction in the present paper.

Niiniluoto [18] observed that the branch of applied mathematics that studies inverse problems deals with abductive types of inference. This proposal needs serious qualifications. Niiniluoto takes abduction to be reasoning from effects to causes. This is at the same time a limiting view of abduction and a too coarse-grained one. It is limitative, as it implies that some (e.g., well-posed, continuous and parametric-model) inverse problems, which in reality are largely the matter of deductive inferences, would become instances of abductive inference. For example, in *well-posed*

---

[7]For example, Peirce recommends relying on early guesses, those that are *"gravid* with young truth"* (EP 2:472), the notion which he had gleaned from his studies on the logic of the history of scientific discoveries. Why this works is that it expedites the inquiry that sooner or later would have! discovered those conjectures.

inverse problems, where the relevant parameters or properties of models are known so that the solution depends continuously on the available data, the predominant mode of reasoning is deductive rather than abductive one.

Niiniluoto's proposal is too coarse, on the other hand, as conceiving abduction as an inverse reasoning from effects to causes masks the details of what is going on in scientific reasoning and even may misleadingly suggest that abduction has such explanatory virtues as confirmatory degrees of beliefs in the likelihood of the hypotheses.

Peirce's interrogative construal, however, suggests both a broader and a more specific view of abduction. It is broader, as it is fitted also for situations in which those strict cause-effect relationships that Niiniluoto requires may be unobtainable. Those situations typically concern problem contexts that are severely understructured. There are endless examples of such contexts in the sciences. In the area of inverse problems, for example, such contexts give rise to *ill-posed* problems, such as those where the time-reversibility fails when the converse of a continuous mapping is a discontinuous one and so that analog samples no longer work, or where the models are non-parametric, where there is too much noise, or where there is only some anecdotal data at hand, and so on. These are very common situations in contemporary science and fieldwork. Yet the investigation must go on. What would a scientist do? Examples such as these cry out for abduction in its interrogative or investigand mood, where the scientists are "guessing at the unknown unknowns" [26]. For example, when confidence regions or choosing good statistical parameters that do not 'under-smooth' are at issue, a statistician typically makes a guess. But such guesses are not blind guesses. They are guided by a relatively precise understanding of the kind of uncertainty and ignorance involved in the situation. Examples of such types of abductive reasoning in contemporary sciences are numerous, and have been explored in [27].

3   A Preliminary Analysis of Peirce's Abduction

In order to avoid abduction from deteriorating into deductive, over-simplified or static schemas, which would appear more like black boxes from evidence, effects or conclusions to hypotheses, we develop here a first, preliminary analysis of what the dynamic processes mean in reasoning in general.

Logical reasoning can be viewed as a dynamic process of information flow from premises to the conclusion. The notion of information and its content in logic has been studied in [4]. We first explain the information flow in a deductive reasoning, and then extend this perspective to Peirce's abductive reasoning. This, we emphasise, is a *preliminary* analysis following a straightforward interpretation of Peirce's abduction using classical truth-value configurations of propositions. Although it quickly turns out that this first analysis is much too simple to capture the meaning of scientific inquiry or the making of plausible conjectures of the abductive kind, we can learn the basic idea of logical dynamics of information flow from these remarks.

Consider the Modus Tollens (MT for short). Assume that you know that C is true if A is true, and you learn that C is not true. Then logic tells you that A is not true. Formally, we write

$$A \to C, \neg C \vdash \neg A.$$

What is the information flow in this piece of deductive reasoning? Intuitively, the information provided by the premises is a *range of states* which tells you the possibilities given by the truth values of propositions. Let $\mathbf{T}$ be the truth value TRUE, and $\mathbf{F}$ be the truth value FALSE. The rule of MT involves two propositions, and hence there are only four possibilities regarding their truth values. This is represented in Table 2.

TT	**TF**
FT	**FF**

Table 2: Range I

We call each possibility a *state*. Every state consists of a truth value of A and a truth value of C. The range of states represents the basic information of MT, that is, all the truth-value configurations for propositions that occur in MT. Identifying the range of states is not a part of the MT reasoning. When the reasoning is started, the information concerning the range of states will be changed as the premises are being fed in.

The major premise of the MT is the proposition $A \to C$. It tells us that C is true if A is true, that is, the state where A is true but C is false does not occur. This is the logical information contained in the implication. Now we start the reasoning in MT. Input $A \to C$. The range I in Table 2 is changed into the range II in Table 3.

TT	—
FT	**FF**

Table 3: Range II

The state **TF** is removed from the range I by the new information $A \to C$. Then we input the minor premise $\neg C$ which means that $\neg C$ is true. The information involved in $\neg C$ is that the opposite truth value is obtained if we give a truth value for C. Hence C is false. And hence the states **TT** and **FT** cannot survive since they allow only **T** for C. Finally, we obtain the final range of states in Table 4.

$$\boxed{\begin{array}{cc} - & - \\ - & \textbf{FF} \end{array}}$$

Table 4: Range III

Let us call this elimination process the *dynamic process of the range of information*. It is seen to provide a rough characterization of what deductive reasoning means:

- A pattern of propositional reasoning P is deductive, if and only if there is no uncertainty at the final stage of the dynamic process of the range of information on P.

Note that the uncertainty concerns not the truth values per se but whether we are in the position to determine the conclusions's truth value or not. Naturally, there is no state at the final stage of the dynamic process in which the conclusion in P would come out as false. In other words, deductive reasoning is maximally *secure*: it approaches certainty to the fullest degree when compared with other, non-deductive kinds of reasoning. But of course the conclusion in deductive reasoning can never be any more certain than any of its premises.

As in all reasoning, information flows in deductive reasoning. Information as a range provides an appropriate basis for such dynamics. Every pattern of deductive reasoning in propositional logic can be interpreted as a dynamic process exhibited in the range of information.

Let us now consider Peirce's abduction that has an interrogative conclusion as well as a negated minor premise. One direct formalization of it is to use classical propositional logic with a question mark in the conclusion. We can have the following forms to be contrasted with MT:

$$A \to C, C \vdash ?\neg A. \tag{1}$$

$$A \to \neg C, \neg C \vdash ?\neg A. \tag{2}$$

Given the classical reading of negation, these two patterns are essentially the same. Peirce's 1905 formulation of abduction took both to correspond to MT. The key ingredient is the question mark ?, which represents the interrogative mood in the conclusion. Recall that the conclusion $?\neg A$ is to indicate that 'it is to be inquired whether A is not true'.[8]

With respect to (1), using our analysis of information flow as a MT we have the following dynamic process of changes in the range of information:

$$
\begin{array}{|cc|}
\hline
\textbf{TT} & \textbf{TF} \\
\textbf{FT} & \textbf{FF} \\
\hline
\end{array}
\overset{A \to C}{\Longrightarrow}
\begin{array}{|cc|}
\hline
\textbf{TT} & - \\
\textbf{FT} & \textbf{FF} \\
\hline
\end{array}
\overset{C}{\Longrightarrow}
\begin{array}{|cc|}
\hline
\textbf{TT} & - \\
\textbf{FT} & - \\
\hline
\end{array}
$$

At the final stage, we arrive at the range of two states **TT** and **FT** in which we are *uncertain* whether a definite truth value of A can be established. In Peirce's terms, the reasoning now is *less secure* than the previous one. The uncertainty of this kind is what is contained in the conclusion of abductive reasoning in the interrogative mood ($?\neg A$).

For Peirce, however, less secure reasoning is in a certain quite concrete sense superior to the more secure, deductive type of reasoning: the former is more valuable in its productiveness. Since our hypotheses, models, simulations, scenarios and ultimately theories are bound to be fallible, the value of abduction lies in its power to open up new, future lines of investigation. The trade-off is that when the security decreases, the amount of "uberty" (Peirce's term for those useful and valuable types of productiveness that we find thriving in uncertainty) increases.

Likewise, for (2) we have the following dynamic process:

$$
\begin{array}{|cc|}
\hline
\textbf{TT} & \textbf{TF} \\
\textbf{FT} & \textbf{FF} \\
\hline
\end{array}
\overset{A \to \neg C}{\Longrightarrow}
\begin{array}{|cc|}
\hline
- & \textbf{TF} \\
\textbf{FT} & \textbf{FF} \\
\hline
\end{array}
\overset{\neg C}{\Longrightarrow}
\begin{array}{|cc|}
\hline
- & \textbf{TF} \\
- & \textbf{FF} \\
\hline
\end{array}
$$

The final truth values of A measure the uncertainty that remains in the final stage of the reasoning process.

Peirce's abduction as the MT with an interrogative conclusion can be viewed as reasoning in which the range of information changes along with the input of the premises, achieving a certain measure of uncertainty about the conclusion in the final stage. It is not difficult to see that every abductive reasoning pattern given in classical propositional logic can be analyzed in terms of such ranges of information as shown above.

---

[8]In fact, the negation symbol in Peirce's conclusion $?\neg A$ can be removed, since the uniform pattern for (1) and (2) is $A \to C, C \vdash ?A$. What is significant is the meaning of the question mark.

Moreover, we can generalize the MT-abduction in the following way:

$$A_1 \to C, \ldots, A_n \to C, C \vdash \ ?\neg A_1 \ \& \ \ldots \ \& \ ?\neg A_n \tag{3}$$

$$A_1 \to \neg C, \ldots, A_n \to \neg C, \neg C \vdash \ ?\neg A_1 \ \& \ \ldots \ \& \ ?\neg A_n. \tag{4}$$

In (3) and (4), we have n possible reasons $A_1 \ldots A_n$. The conclusion is the conjunction of interrogative assertions concerning these reasons. Naturally, we could equally well consider disjunctive cases that inquire "whether some of these possibilities are not the case?". Obviously, our analysis on the dynamics of abductive reasoning can be applied to all these patterns.

Up to now we have achieved an interpretation of the interrogative mood in the conclusion $?\neg A$ of abduction; namely, it is uncertainty about the establishment of the truth value of A. However, this obviously is not a sufficient interpretation of what $?\neg A$ means in Peirce's novel 1905 schema of abduction. It suffices to recall that the starting point of abductive stage of inquiry is the observed surprising fact, and the end of abduction is a formulation of a conjecture submitted to investigation. Intuitively, $?\neg A$ means that it is to be inquired whether $\neg A$ is a plausible conjecture or not. Uncertainty of course cannot do the whole job of living inquiry. It only nurtures it by increasing the degrees of freedom for the lines of research scientific projects can undertake. The inquiry comes to its full blossom only after satisfactory interrogative conclusions have been made. But then the obvious question that arises is: what is the model for selecting plausible reasons for a surprising fact among so many other reasons? Even further, is there, at least in an ideal case, a unique, most plausible such reason? To address these questions, we need to seek and find answers to $?\neg A_1, \ldots, ?\neg A_n$ just as the schemas (3) and (4) invite us to do.

Given a surprising fact F, there may be many, or even an endless number of, potential propositions that could count as possible reasons that could explain how F ceases to be surprising. In order to form *plausible* conjectures, on the other hand, we need to state which propositions are within the range of such plausible ones. Even more specifically, we might need to find a way to state which ones among the full spectrum of plausible conjectures are more plausible than others, or even further, which among the more plausible ones may be counted among the most plausible such hypotheses. In the following sections, we will use neighborhood models and their updates in order to achieve a better logical grasp of how inquirers may find plausible conjectures. The questions concerning establishment of hierarchies or orderings between a number of plausible conjectures is left for a further occasion.

4 A Logic of Conjecture

In this section, we introduce a logic of conjecture which is interpreted in neighborhood models. This logic describes the relationship between the occurrence of a surprising fact and the investigand, namely the reason to initiate inquiry that could ascertain the plausibility of the conjecture. After a surprising fact F occurs to an agent, some possible propositions that might support F occur to the agent which we call conjectures. It is the surprising facts that force the agents to produce such candidates for possible reasons to account for those facts.

To develop a logic of conjecture we need a formal language to express that "the agent *conjectures that*" A is true. A modal epistemic language that is widely used in (dynamic) epistemic logic will come in handy.

The desired language $\mathcal{L}_{\mathrm{EL}}$ consists of a denumerable set $\mathcal{V}$ of propositional variables, propositional connectives $\neg$ (negation), $\vee$ (disjunction) and the modal attitude operator $\square$. The set of all formulas is defined inductively as follows:

$$A ::= p \mid \neg A \mid (A \vee A) \mid \square A,$$

where $p \in \mathcal{V}$. Other propositional connectives $\wedge$ (conjunction) and $\rightarrow$ (implication) can be defined as usual. We read the formula $\square A$ as that the agent *conjectures that* A is true, while its dual $\diamond A$, that the agent does not conjecture that A is false, may be read as that the agent reckons the conjecture A possible. (That is, "it is possible, for all that the agent can conjecture, that A".)[9]

---

[9]Instead of conjecturing, the modal attitude involved in Peirce's abductive reasoning could well be be taken to be that of *hope*, in the qualified sense of the *scientific hope* on our or future generations' powers to conjecture or guess right. Thus we could also read the modal operator as "the agent *hopes* that A would be the case". Evidence for 'hope' as the central attitude involved in abduction comes from Peirce's letter to F. A. Woods (MS L 477, 1913): "I have always, since early in the sixties, recognized three different types of reasoning, viz.: First, *Deduction* which depends on our confidence in our ability to analyze the meanings of the signs in or by which we think; second, *Induction*, which depends upon our confidence that a run of one kind of experience will not be changed or cease without some indication before it ceases; and third, *Retroduction*, or Hypothetic Inference, which depends on our *hope, sooner or later, to guess at the conditions under which a given kind of phenomena will present itself*" (added emphasis). Hope, we believe, is an all-important and neglected modal attitude, and even more important than belief, in scientific reasoning and discovery. It does not seem to have been studied before in philosophy of science or in modal logics. In logic it would give rise to a new class of '*spestic*' logics. In philosophy of science, it would replace the doxastic approaches that are either too far removed from the actual practice of scientists or else in a conceptual deadlock when founded on the orthodox Bayesian reasoning. To have 'hope' in the truth of propositions comes moreover closer in spirit to Peirce's limit notion of truth and his logic of inquiry than belief does. Far from meaning 'faith', hope is an expression of a genuine (i.e., non-sham and non-fake) scientific attitude which meets fundamental uncertainty concerning future data, and where beliefs or degrees of belief in the hypothesis would come out as much too strong.

Remark 1. One may wonder about the meaning of the question mark in our notation ?¬A in Section 3 that formalizes the interrogative conclusion. As commented above, the interrogative conclusion is not merely a question that asks weather A is true or not. The meaning of ?¬A cannot be identified with the meaning of a pure question as they occur in natural language. There are indeed many erotetic logics of question in the literature. For example, [32] take asking a question as drawing an equivalence relation in the epistemic Hintikka-Kripke model. But we cannot see how such interpretations could be applied to abduction at all. A similar comment applies to the suitability of inquisitive semantics [7] for abductive problems, as while perhaps mildly pragmatistic in its motivation, the models of inquisitive logic fail to reach the generality of our neighbourhood models. In fact, we intend to interpreted ?¬A as a determination to find out whether the conjecture A is plausible or not. It is a request for finding new data, not unlike those questions that can be interpreted as requests put upon the sources of information to bring it about whether something is the case. The proposition that A is a conjecture is expressed by the formula $\Box A$. Hence ?¬A means that $\Box A$ is either plausible or not. The presupposition of such a 'question' is of course readily established. In the abduction $A \to C, C \vdash ?¬A$, the proposition C is the surprising fact, and the premiss $A \to C$ suggests A to acquire the status of a conjecture concerning novel findings. But it is the *plausibility* of A that really is in question in our framework.

The models for the language $\mathcal{L}_{\text{EL}}$ that we use are neighborhood models. They are used in non-normal modal logics (cf. [6]) to represent the information range and also do so in our analysis of the dynamic process of abduction. We recall some basic definitions which can be found in [17].

Definition 2. A *neighborhood frame* is a pair $\mathfrak{F} = (W, \tau)$, where W is a non-empty set and $\tau : W \to \mathcal{P}(\mathcal{P}(W))$ is a function from W to the power set of the power set of W. For each $w \in W$, a set $X \in \tau(w)$ is called a *neighborhood* of w. A *neighborhood model* is a structure $\mathfrak{M} = (W, \tau, V)$, where (W, τ) is a neighborhood frame and $V : \mathcal{V} \to \mathcal{P}(W)$ is a valuation.

Given a neighborhood model $\mathfrak{M} = (W, \tau, V)$ and $w \in W$, we may think of points in W as the possible worlds or *mental states* of the inquirer or the group of scientists. The set $\tau(w)$ represents the set of all possible propositions subject to the scientific attitudes of the agent, namely conjectures that the agent or the team may make at the current world w. Using such neighborhood models one can easily update them by deleting propositions which are not plausible. This is the advantage of our approach that we take, as compared to the standard Kripke models commonly used in epistemic logic.

88

Definition 3. The *truth* of a formula A at a world w in a model $\mathfrak{M}$ (notation: $\mathfrak{M}, w \models A$) can be defined recursively as follows:

$\mathfrak{M}, w \models p$ iff $w \in V(p)$.
$\mathfrak{M}, w \models \neg A$ iff $\mathfrak{M}, w \not\models A$.
$\mathfrak{M}, w \models A \vee B$ iff $\mathfrak{M}, w \models A$ or $\mathfrak{M}, w \models B$.
$\mathfrak{M}, w \models \Box A$ iff $[\![A]\!]_\mathfrak{M} \in \tau(w)$, where $[\![A]\!]_\mathfrak{M} = \{u \in W \mid \mathfrak{M}, u \models A\}$.

We say that a formula A is *valid*, if $[\![A]\!]_\mathfrak{M} = W$ for every neighborhood model $\mathfrak{M} = (W, \tau, V)$.

We say that a neighborhood frame $\mathfrak{F} = (W, \tau)$ is *monotone* when for all $w \in W$, if $X \in \tau(w)$ and $X \subseteq Y \subseteq W$, then $Y \in \tau(w)$. Naturally the same applies to the models, too. Over monotone neighborhood models, we rewrite the semantic clause for conjecturing as:

$$\mathfrak{M}, w \models \Box A \text{ iff there exists } X \in \tau(w) \text{ such that } X \subseteq [\![A]\!]_\mathfrak{M}.$$

The attitude logic considered here is the set of all formulas valid over all neighborhood models, which is finitely axiomatizable as the modal logic **E** in [6] is. That is, it is the smallest set of modal formulas containing the classical propositional logic which is closed under the following rules:

(MP): from A and $A \rightarrow B$ infer B.
(E): from $A \leftrightarrow B$ infer $\Box A \leftrightarrow \Box B$.

Moreover, the modal logic **M** which is the extension of **E** that adds the monotonicity rule (M): from $A \rightarrow B$ infer $\Box A \rightarrow \Box B$, is sound and complete with respect to the class of all monotone neighborhood models (cf. [6]).

Next, we analyze the updates of a neighborhood model and apply that analysis to the dynamic process of abduction. Given a neighborhood model $\mathfrak{M} = (W, \tau, V)$ and $w \in W$, the neighborhoods of w are in the set $\tau(w)$. Consider now a surprising fact F appearing to the agent. The $\tau(w)$ will be changed when the agent asks which propositions in $\tau(w)$ are plausible conjectures for F. Note that the denotation of F in $\mathfrak{M}$ is its truth set $[\![F]\!]_\mathfrak{M}$. The update can be described as follows:

If a neighborhood $X \in \tau(w)$ *supports* F, then keep F in the updated model. Otherwise, delete X from $\tau(w)$ in the updated model.

We need to explain what supporting means. Formally, a neighborhood $X \in \tau(w)$ *supports* F if and only if $X \subseteq [\![F]\!]_\mathfrak{M}$. Then we can obtain the following formal definition of the update of a neighborhood model (cf. [17]):

Definition 4. Given a neighborhood model $\mathfrak{M} = (W, \tau, V)$ and a formula F, the updated model $\mathfrak{M}^F = (\llbracket F \rrbracket_\mathfrak{M}, \tau^F, V^F)$ of $\mathfrak{M}$ by F is defined as follows:

$\tau^F(w) = \{X \subseteq \llbracket F \rrbracket_\mathfrak{M} \mid X \in \tau(w)\}$, for every $w \in \llbracket F \rrbracket_\mathfrak{M}$.
$V^F(p) = V(p) \cap \llbracket F \rrbracket_\mathfrak{M}$, for every $p \in \mathcal{V}$.

Note that the updated model is monotone if $\mathfrak{M}$ is monotone: suppose $X \in \tau^F(w)$ and $X \subseteq Y \subseteq \llbracket F \rrbracket_\mathfrak{M}$. Then $X \in \tau(w)$. Hence $Y \in \tau(w)$ by the monotonicity of τ. Therefore, $Y \in \tau^F(w)$.

To support a surprising fact F is to have at least one conceivable situation, namely the truth set generated by that fact. To have an updated neighborhood $\tau^F(w)$ is exactly to have the range of propositions which support F.

5 Two Examples

We present two examples from the classical literature that illustrate the dynamics of the logic of abduction.

Example 5. First, we analyze the following example in the context of commonplace reasoning given by Peirce's student John Dewey [8]:

> Projecting nearly horizontally from the upper deck of the ferryboat on which I daily cross the river, is a long white pole, bearing a gilded ball at its tip. It suggested a flagpole when I first saw it; its color, shape, and gilded ball agreed with this idea, and these reasons seemed to justify me in this belief. But soon difficulties presented themselves. The pole was nearly horizontal, an unusual position for a flagpole; in the next place, there was no pulley, ring, or cord by which to attach a flag; finally, there were elsewhere two vertical staffs from which flags were occasionally flown. It seemed probable that the pole was not there for flag flying.

> I then tried to imagine all possible purposes of such a pole, and to consider for which of these it was best suited: (a) Possible it was an ornament. But as all the ferryboats and even the tugboats carried like poles, this hypothesis was rejected. (b) Possible it was the terminal of a wireless telegraph. But the same considerations made this improbable. Besides, the more natural place for such a terminal would be the highest part of the boat, on top of the pilot house. (c) Its purpose might be to point out the direction in which the boat is moving.

> In support of this conclusion, I discovered that the pole was lower than the pilot house, so that the steersman could easily see it. Moreover, the tip was enough higher than the base, so that, from the pilots's position, it must appear to project far out in front of the boat. Moreover, the pilot being near the front

of the boat, he would need some such guide as to its direction. Tugboats would
also need poles for such a purpose. This hypothesis was so much more probable
than the others that I accepted it. I formed the conclusion that the pole was set
up for the purpose of showing the pilot the direction in which the boat pointed,
to enable him to steer correctly. ([8, pp. 70–71])

Dewey's example relates to familiar, common sense observations. However, one is
using abduction in ordinary life much in the same sense as in complex scientific
inquiry. At the beginning, the surprising fact is the following proposition:

F: There is a long white pole bearing a gilded ball at its tip and near to the upper
deck of the ferryboat.

Then the following four investigands are considered to be possible, consistent hypotheses:

A_1: Is it not that the long white pole is a flagpole?

A_2: Is it not an ornament?

A_3: Is it not the terminal of a wireless telegraph?

A_4: Is it not the pole for pointing out the direction in which the boat is moving?

All these hypotheses support F at some stage. It is the proposition F that makes
the agent to suggest possible reasons A_1 to A_4. The surprising fact eliminates other
unrelated propositions about the fact. For example, the fact that the pole was made
by a certain company is irrelevant to the situation at the current state that the
agent wants to know the function of the pole. Thus we achieve a stage which can
be represented as the following neighborhoods of the current state:

A_1	A_2
A_3	A_4

Some of these neighborhoods are dropped when some more surprising facts are
observed. Consider the following facts appearing in the example:

C_1: There are similar flagpoles on other places.

C_2: Every boat has a pole like this in the same position.

C_3: Similar boats carried similar poles, and a better place for the terminal of a
wireless telegraph would be in a higher position.

C_4: The pole was lower than the pilot house.

When the fact C_1 is suggested, A_1 does not support it. Hence A_1 is deleted from the neighborhoods. Similarly after C_2 and C_3 are suggested, A_2 and A_3 are deleted. Finally, A_4 supports facts like C_4, and it is thus kept in the neighborhoods and identified as the only remaining explanation for the function of the pole.

Dewey's example conveys well his insight that the investigand has the status of a *working hypothesis*: a preliminary, incomplete conjecture whose plausibility can change as the inquiry goes on.[10]

Example 6. Our second example is the famous Kepler case, namely Kepler's discovery of the Martian orbit being elliptical. This was Peirce's favourite argument for the reality of abductive reasoning in the history of science.

> For example, at a certain stage of Kepler's eternal exemplar of scientific reasoning, he found that the observed longitudes of Mars, which he had long tried in vain to get fitted with an orbit, were (within the possible limits of error of the observations) such as they would be if Mars moved in an ellipse. The facts were thus, in so far, a *likeness* of those of motion in an elliptic orbit. Kepler did not conclude from this that the orbit really was an ellipse; but it did incline him to that idea so much as to decide him to undertake to ascertain whether virtual predictions about the latitudes and parallaxes based on this hypothesis would be verified or not. This probational adoption of the hypothesis was an Abduction. (CP 2.96, c.1902)

After Peirce, the details of Kepler's five-year process of discovery have been discussed abundantly by the historians and philosophers of science (see e.g. [15, 16, 35]).

Here we relate those steps that lead to Kepler's discovery to the dynamic process of abduction.

At the beginning of Kepler's quest, the surprising fact is the following proposition:

F: The planet Mars has non-uniform speed.

Years of painstaking investigation had Kepler consider the following four propositions at the various stages of his inquiry as those plausible hypotheses consistent with the fact F:

A_1: Is not the orbit of Mars circular in the geometric sense? (This is Kepler's *Vicarious Hypothesis*, the goal of which was to develop a new mathematical model by which to calculate Mars's longitudinal positions.)

---

[10]We thank Sami Paavola for this remark.

A_2: Is not the orbit of Mars circular and epicyclic in the physical sense? (This is Kepler's *Area Method*, the goal of which was to replace the geometric equant point (A_1) by two further ideas: the area principle and the epicycle theory that describe the real physical forces.)

A_3: Is not the orbit of Mars oval shaped?

A_4: Is not the orbit of Mars an ellipse?

All these hypotheses support F at some stage. A_1 supports F in its geometric structure. A_2 supports F as regards to the behavior of physical forces. A_3 supports the fact F, since the oval-shaped curve is an approximation of the circle in the quadrants as well as the observed errors in the octants. The fact is being supported by A_4, namely the elliptical orbit, too, as elliptical shapes are mathematical approximations to oval-shaped orbits.

More facts were then observed along Kepler's investigation:

C_1: The equant point makes the angular velocity of Mars around it uniform. Therefore, the planet's speed along its path is also non-uniform, though according to the Vicarious Hypothesis, the equant point is not a physically real point in space.

C_2: There is no acceptable model of Mars to be derived from the area method, as the calculated longitudes do not fully agree with A_1.

C_3: The calculated longitudes do not fully agree with A_3 with an auxiliary ellipse that approximates oval-shaped orbit.

C_4: Observations and implications of the physical theory can be saved only if the orbit of Mars is an ellipse.

Since no purely geometric description of the celestial movement was satisfactory for Kepler, A_1 fails to support C_1. A_2 fails to support C_2 due to the absence of any observed actual physical fact to imply the hypothesis. (It is at this point of encountering the observed errors in the longitudinal measurements that Kepler in fact began to doubt the circular orbit assumption, in other words his finding of the famous 8-minute discrepancy in the longitudinal measurement in the octants, in modern terms the eccentricity in the orbital motion of planets.) Finally, A_3 fails to support C_3, as calculating the longitudes with an auxiliary ellipse also resulted in similar reversed errors in the octants.

We are therefore left with A_4, the elliptical orbit hypothesis. The fact C_4, it needs to be remarked, is supported by A_4 in the sense that—unlike is the case with

A_1—A_4 does not require any mathematical and geometrical account that has no known relation to reality (known to Kepler at the time of his inquiry, that is). So it is A_4 that is kept in the neighborhoods and taken as the sole remaining explanation for the surprising fact observed concerning the non-uniform movement of Mars.

In this example, Peirce assigns abductive conclusions the status of probationary hypotheses much like Dewey's working hypotheses later on were intended to capture. Those hypotheses are invitations to further inquiry, which is what the investigand mood aims to convey.

6   A Logic of Abduction

In Section 4 we defined the logic of making conjectures, in which no dynamic mechanism is introduced. In this section, we present the logic of abduction that we desire, which is the dynamic extension of the logic of conjecture. This logic was presented in [17] in the setting of monotone neighborhood models. Here we represent it in the class of all neighborhood models, as that is what the abductive reasoning modes really call for.

The language of the dynamic conjecture logic $\mathcal{L}_{DC}$ is obtained from $\mathcal{L}_{EL}$ by adding dynamic operators of the form $[!F]$ where F is a formula. The set of $\mathcal{L}_{DC}$-formulas is defined recursively as follows:

$$A ::= p \mid \neg A \mid (A \vee A) \mid \Box A \mid [!A]A,$$

where $p \in \mathcal{V}$. Define $\langle !A \rangle B := \neg[!A]\neg B$. Intuitively, $[!F]A$ means that the hypothesis A becomes true after the occurrence of the surprising fact F.[11]

In this way we can read the formula $[!F]\Box A$ as follows: Agent's conjecturing or hoping for A to turn out to be a plausible one is survived in the updated neighborhood of the current state. In other words, it is not only the proposition but the *attitude* the agent or the team of inquirers have towards the status of the proposition A that survives in the face of some surprising fact F. And when that happens, F ceases to be surprising and serves as evidence by virtue of which the abductive hypothesis acquires or increases its plausibility.

---

[11] Or, alternatively but not equivalently, the hypothesis remains unfalsified. The two are not equivalent, since while the first affirms that A is the case, the second, Popper-style update only states that the hypothesis has not been proved to be false. We thank the reviewer for raising the point that the first interpretation may be seen as the natural one, where A becomes the case after the occurrence of the surprising fact, whereas the second deals with whether the hypothesis (conjecture) A has been discarded or not, something that has to do with whether A is found in the agent's neighbourhood.

Definition 7. Let $\mathfrak{M} = (W, \tau, V)$ be a neighborhood model. The *truth* of an $\mathcal{L}_{\mathrm{DC}}$-formula A (notation: $\mathfrak{M}, w \models A$) is defined recursively as the semantics for $\mathcal{L}_{\mathrm{S}}$, adding the following semantic clause:

$$\mathfrak{M}, w \models [!F]A \text{ iff } \mathfrak{M}, w \models F \text{ implies } \mathfrak{M}^F, w \models A.$$

The logic **DC** is defined as the set of all $\mathcal{L}_{\mathrm{DC}}$-formulas valid over the class of all neighborhood models. The logic **DCM** is the set of all $\mathcal{L}_{\mathrm{DC}}$-formulas valid over the class of all monotone neighborhood models.

These two logics are finitely axiomatizable, by adding the reduction axioms (R_1–R_5) in Table 5 to **E** and **M** respectively.

(R_1)	$[!F]p \leftrightarrow (F \to p)$
(R_2)	$[!F]\neg A \leftrightarrow (F \to \neg[!F]A)$
(R_3)	$[!F](A \vee B) \leftrightarrow [!F]A \vee [!F]B$
(R_4)	$[!F]\Box A \leftrightarrow (F \to \Box \langle !F \rangle A)$
(R_5)	$[!F][!G]A \leftrightarrow [!(F \wedge [!F]G)]B$

Table 5: Reduction Axioms

Theorem 8. *The logic* **DC** *(***DCM***) is sound and complete with respect to the class of all (monotone) neighborhood models.*

Proof. Here we sketch the proof (cf. [17]). Assume that $A \notin \mathbf{DC}$. By the reduction axioms R_1 to R_5, A is equivalent to a formula $A^* \in \mathcal{L}_{\mathrm{EL}}$, i.e., without occurrences of dynamic operators. Then $A^* \notin \mathbf{E}$, for otherwise A would be a theorem of **DC** since **DC** is an extension of **E**. By the completeness of **E**, there exists a neighborhood model $\mathfrak{M}$ and a state w such that $\mathfrak{M}, w \not\models A^*$. Then $\mathfrak{M}, w \not\models A$. Hence the completeness is obtained. For the soundness, it suffices to verify the validity of those reduction axioms. Here we recall only the validity of the reduction axiom $[!F]\Box A \leftrightarrow (F \to \Box \langle !F \rangle A)$. Assume $\mathfrak{M}, w \models [!F]\Box A$ and $\mathfrak{M}, w \models F$. Hence $\mathfrak{M}^F, w \models \Box A$. Then $\llbracket A \rrbracket_{\mathfrak{M}^F} \in \tau^F(w)$. Hence $\llbracket A \rrbracket_{\mathfrak{M}^F} \subseteq \llbracket F \rrbracket_{\mathfrak{M}}$ and $\llbracket A \rrbracket_{\mathfrak{M}^F} \in \tau(w)$. It is easy to check that $\llbracket A \rrbracket_{\mathfrak{M}^F} = \llbracket F \wedge [!F]A \rrbracket_{\mathfrak{M}} = \llbracket \langle !F \rangle A \rrbracket_{\mathfrak{M}} \in \tau(w)$. Hence $\mathfrak{M}, w \models \Box \langle !F \rangle A$. Conversely, assume $\mathfrak{M}, w \models F \to \Box \langle !F \rangle A$ and $\mathfrak{M}, w \models F$. Then $\mathfrak{M}, w \models \Box \langle !F \rangle A$. Hence $\llbracket \langle !F \rangle A \rrbracket_{\mathfrak{M}} \in \tau(w)$. Then $\llbracket A \rrbracket_{\mathfrak{M}^F} \in \tau(w)$. Since $\llbracket A \rrbracket_{\mathfrak{M}^F} \subseteq \llbracket F \rrbracket_{\mathfrak{M}}$, we have $\llbracket A \rrbracket_{\mathfrak{M}^F} \in \tau^F(w)$. Hence $\mathfrak{M}^F, w \models \Box A$. Therefore $\mathfrak{M}, w \models [!F]\Box A$. The case for the logic **DCM** is quite similar. $\quad\square$

In the updated model $\mathfrak{M}^F = (\llbracket F \rrbracket_{\mathfrak{M}}, \tau^F, V^F)$ of a given neighborhood model $\mathfrak{M}$ by a surprising fact F, the support relation is involved: a neighborhood $X \in \tau(w)$ survives in $\tau^F(w)$ if and only if X supports F. This fact allows us to introduce a general logic for support. We introduce a binary operator $\prec$ for the support relation. We read $A \prec B$ as that the proposition A *supports* B in the sense of turning B into a plausible conjecture.

The language $\mathcal{L}_S$ of the support logic **S** consists of a denumerable set $\mathcal{V}$ of propositional variables, propositional connectives $\neg$ and $\vee$, and the binary supporting operator $\prec$. The set $\mathcal{L}_S$ of all formulas is defined recursively as follows:

$$A ::= p \mid \neg A \mid (A \vee A) \mid (A \prec A) \mid \mathsf{U}A,$$

where $p \in \mathcal{V}$. Other propositional connectives $(\top, \bot, \wedge, \rightarrow$ and $\leftrightarrow)$ are defined as usual. Define $\Box A := A \prec \top$ and $\Diamond A := \neg\Box\neg A$. The unary operator U is the global modality. We read $\mathsf{U}A$ as that A holds everywhere.

The semantics for $\mathcal{L}_S$ will be given in arbitrary neighborhood models (and so it is not restricted to monotone ones).

Definition 9. Given a neighborhood model $\mathfrak{M} = (W, \tau, V)$ and $w \in W$, the *truth* of a formula A at a state w in $\mathfrak{M}$ (notation: $\mathfrak{M}, w \models A$) is defined recursively as follows:

$\mathfrak{M}, w \models p$ iff $w \in V(p)$
$\mathfrak{M}, w \models \neg A$ iff $\mathfrak{M}, w \not\models A$
$\mathfrak{M}, w \models A \vee B$ iff $\mathfrak{M}, w \models A$ or $\mathfrak{M}, w \models B$
$\mathfrak{M}, w \models A \prec B$ iff $\llbracket A \rrbracket_{\mathfrak{M}} \in \tau(w)$ and $\llbracket A \rrbracket_{\mathfrak{M}} \subseteq \llbracket B \rrbracket_{\mathfrak{M}}$
$\mathfrak{M}, w \models \mathsf{U}A$ iff $\llbracket A \rrbracket_{\mathfrak{M}} = W$,

where $\llbracket A \rrbracket_{\mathfrak{M}} = \{u \in W \mid \mathfrak{M}, u \models A\}$ and $\llbracket B \rrbracket_{\mathfrak{M}} = \{v \in W \mid \mathfrak{M}, v \models B\}$ are the truth sets of A and B in $\mathfrak{M}$. We say a formula A is *valid*, if $\mathfrak{M}, w \models A$ for every neighborhood model $\mathfrak{M}$ and state w in $\mathfrak{M}$.

It follows from the semantics directly that the formula

$$(A \prec B) \leftrightarrow \Box A \wedge \mathsf{U}(A \rightarrow B)$$

is in **S**. Moreover, by the definition, we have

$$\mathfrak{M}, w \models \Box A \text{ iff } \llbracket A \rrbracket_{\mathfrak{M}} \in \tau(w).$$

Apparently, this coincides with the semantics of $\Box$ as given in Definition 3.

The support logic **S** is defined as the set of all valid formulas in our semantics.[12] Now the task is to explore the dynamic extension of **S**. The language $\mathcal{L}_{DS}$ of the dynamic extension is obtained from $\mathcal{L}_S$ by adding dynamic operators of the form $[!F]A$. We use **DS** to denote the dynamic logic of support. For getting a complete axiomatization of **DS**, it remains to work out the reduction axioms for $\prec$ and U.

Theorem 10. *The following reduction axioms are valid:*

(R_6) $[!F]\mathsf{U}A \leftrightarrow (F \to \mathsf{U}[!F]A)$.

(R_7) $[!F](A \prec B) \leftrightarrow (F \to \Box\langle !F\rangle A \wedge [!F]\mathsf{U}(A \to B))$.

Proof. For (R_6), assume $\mathfrak{M}, w \models [!F]\mathsf{U}$ and $\mathfrak{M}, w \models F$. Then $\mathfrak{M}^F, w \models \mathsf{U}A$. Let v be any state in $\mathfrak{M}$. Assume $\mathfrak{M}, v \models F$. Then $v \in [\![F]\!]_{\mathfrak{M}}$. Since $\mathfrak{M}^F, w \models \mathsf{U}A$, we have $\mathfrak{M}^F, v \models A$. Hence $\mathfrak{M}, v \models [!F]A$. Then $\mathfrak{M}, w \models \mathsf{U}[!F]A$. Conversely, assume $\mathfrak{M}, w \models F \to \mathsf{U}[!F]A$ and $\mathfrak{M}, w \models F$. Then $\mathfrak{M}, w \models \mathsf{U}[!F]A$. Let v be any state in $\mathfrak{M}^F$. Then $\mathfrak{M}, v \models F$ and $\mathfrak{M}, v \models [!F]A$. So $\mathfrak{M}^F, v \models A$. Hence $\mathfrak{M}^F, w \models \mathsf{U}A$.

For (R_7), since $(A \prec B) \leftrightarrow \Box A \wedge \mathsf{U}(A \to B) \in \mathbf{S}$, by reduction axioms ($R_5$) and ($R_6$), we get ($R_7$). $\square$

The reduction axioms reveal how the relation of supporting is affected by updates as they reduce dynamic formulas to static formulas.

Theorem 11. *The dynamic logic of support* **DS** *is axiomatized by all theorems of* **S** *and reduction axioms* (R_1)–(R_5) *in Table 5 and* (R_6) *in Theorem 10.*

Proof. The proof is quite similar to the proof of Theorem 8. $\square$

7   The Two Examples Revisited

It remains to be checked how **DC** serves as the logic of abduction in some concrete cases. To do so, we consider the two examples given in Section 5.

Example 5*. Let $\mathfrak{M} = (W, \tau, V)$ be a neighborhood model and $w \in W$ satisfying the following conditions:

- $\tau(w) = \{V(A_i) \mid V(A_i) \neq \emptyset\ \&\ 1 \leq i \leq 4\}$;

---

[12]The logic **S** may be axiomatized by adding, first, the formula $(A \prec B) \leftrightarrow \Box A \wedge \mathsf{U}(A \to B)$ into **S** and, second, adding additional axioms for the universal modality U to the logic **E**. The global modality is introduced in order to add the reduction axiom (R_7) in Theorem 10. The classical modal logic with universal modality is studied systematically in [10]. How to use global modalities in neighborhood models is currently unknown. This problem will be addressed in another work. Here we concentrate on the logic **S** defined in the semantic way.

- $V(C_i) \neq \emptyset$ for $1 \leq i \leq 4$.

During the abductive process of investigation, the agent who observed the ferryboat's device made a series of conjectures $A_1, \ldots, A_4$. Then $\mathcal{M}, w \models \Box A_i$ for $1 \leq i \leq 4$. This is verified by the fact that $V(A_i) \in \tau(w)$. When the first surprising fact C_1 occurs, we have that $\mathfrak{M}, w \not\models A_1 \prec C_1$. This is determined by the agent's knowledge on the relationship between the surprising fact C_1 and the conjecture A_1. Hence the model $\mathfrak{M}$ is updated by C_1 and we get the new model $\mathfrak{M}^{C_1} = (V(C_1), \tau^{C_1}, V^{C_1})$. Note that $V(A_1) \notin \tau^{C_1}(w)$. Similarly, the model $\mathfrak{M}^{C_1}$ will be updated by the succeeding surprising facts C_2 and C_3. Then we get the updated model $\mathfrak{M}^{C_1, \ldots, C_3}$ in which $V(A_2)$ and $V(A_3)$ are eliminated from $\tau^{C_1, \ldots, C_3}(w)$. At the final stage, A_4 supports C_4. Hence in the model $\mathfrak{M}^{C_1, \ldots, C_3, C_4}$, $V(A_4)$ is the only set in ! the neighborhood of w, and the conjecture A_4 gains plausibility as to the functioning of the pole.

This argument thus demonstrates that $\mathfrak{M}, w \models \langle C_1 \rangle \langle C_2 \rangle \langle C_3 \rangle \langle C_4 \rangle \Box A_4$, that is, A_4 is still conjectured after all the relevant abductive steps have been followed through.

As to the second example, we note that earlier expositions of the Kepler case (notably, by [11]) have indeed argued that Kepler's reasoning is an example of Peirce's 1903 schema of abduction. However, all previous expositions have used Peirce's 1903 schema as if that would have been the definite formulation and the one that could carry out its duties in full. In reality, that schema has nothing to suggest how to take into account the actual dynamics of the process of revision of the model that Kepler was making along the course of his investigation.

Here is how Kepler's reasoning looks like in our dynamic logic of abduction **DC**.

Example 6[*]. Let $\mathfrak{M} = (W, \tau, V)$ be a neighborhood model and $w \in W$ satisfying the following conditions:

- $\tau(w) = \{V(A_i) \mid V(A_i) \neq \emptyset \ \& \ 1 \leq i \leq 4\}$;

- $V(F) \neq \emptyset, V(C_i) \neq \emptyset$ for $1 \leq i \leq 4$.

At the beginning, we have $\mathfrak{M}, w \models A_i \prec F$ for $1 \leq i \leq 4$. Hence in the updated model $\mathfrak{M}^F$, we have $\tau^F(w) = \tau(w) \cap V(F)$. Later, more surprising facts $C_1, \ldots C_4$ occurred in the process of discovery. We have $\mathfrak{M}, w \not\models A_i \prec C_i$ for $1 \leq i \leq 3$, but $\mathfrak{M}, w \models A_4 \prec C_4$. In the process of updating, $V(A_1)$ is eliminated from $\tau^F(w)$, and similarly, $V(A_2)$ and $V(A_3)$ are eliminated from $\tau^{F, C_1}(w)$ and $\tau^{F, C_1, C_3}(w)$. In the final stage, $\tau^{F, C_1, \ldots, C_4}(w) = \{V(A_4)\}$. The argument therefore demonstrates that $\mathfrak{M}, w \models \langle F \rangle \langle C_1 \rangle \langle C_2 \rangle \langle C_3 \rangle \langle C_4 \rangle \Box A_4$, that is, A_4 is the most plausible conjecture after the abductive process has been carried out.

The treatment of the examples is a semantic one. Interestingly, the two examples come out as quite similar. Far from making them redundant, the similarity in question points at an intrinsic connection between the two, which is interesting precisely because they come from two different contexts: the first from a commonsense, everyday observation and the second from a lifelong scientific inquiry of a famous scientist.

Our logic of abduction thus captures those aspects of the interrogative-style abductive reasoning that the 1903 schema masks behind its broad strokes and terse formulation: the dynamics of the investigand that evolves from a tentative, probationary idea or hypothesis towards a plausible conjecture. Such dynamics are part and parcel of actual inquiry, including the Kepler case which for Peirce was an instructive case highlighting the workings of the real discovery in science. But we also find the phenomenon in the Deweyan experimental logic of an everyday reasoner.

8 Discussion

That the logical form of abductive reasoning has in Peirce's work as its conclusion an investigand mood can be likened to that of requests for further information. If this is so, then our findings vindicate Hintikka [14], who proposed that abductive inferences aim at answering the inquirer's questions put to some definite source of information. It remains to be investigated to what extent the abductive logic of support agrees with the interrogative model of inquiry, and how to modify and extend them in order to investigate the suggested allegiance.[13] We expect the interrogative model of inquiry not to fully rationalize all the critical steps of inquiry, however, among them the guessing element at those stages in which uncertainty is in its highest.

Our novel construal of the logic of abduction also vindicates the Gabbay–Woods approach to abduction [9, 36]. Their schema takes abduction to be an ignorance-preserving process that does not advance the epistemic state of the inquirer. Abduction starts from the facts and aims at a plausible conjecture, even when the facts have run out. In our model for abductive logic, a plausible conjecture is just a proposition in the the mind of an agent, or on the draft version of a conference paper by a research team, that can serve as a possible reason for the occurrence of a surprising fact. It need not explain it, although it can well turn out to do so later

---

[13][12] proposes a certain complementarity between dynamic epistemic logic and the interrogative model of inquiry. As often is the case, the mere formal connections, or precisifications, are insufficient to establish the requisite deeper conceptual connections. What is also needed is the reality check that takes those connections through what is done in actual scientific practice and what the experimental condu! ct of the scientists is, including the all-important controlled experiments.

on. Abduction is a process from a surprising fact to the invitation to investigate. New facts come to pass as the investigators look for new sources of information, and old conjectures are deleted from the minds, clouds and preliminary drafts.

Many important issues are left for the future research. For one thing, how does an investigator or a group of investigators gravitate to a limited set of most plausible conjectures? This difficult question cannot be answered as yet. We can can only point out a couple of preliminary suggestions. First, we should avoid the temptation to define the meaning of 'the most plausible conjectures' too simplemindedly as those that are more 'truth-like' or 'increasingly better approximations' to truth. For such approximations are normally measured by prior likelihoods that nevertheless fail to capture essential aspects of scientific practice in discovery. For it is in the nature of scientific inquiry, we believe, that even those hypotheses, models and conjectures that score high on plausibility, are *raw*, that is, they may have to be dumped overnight, if new and surprising evidence, facts, observations and experiments so happen to mandate.

Plausibility orderings are used in [1] to define the concept of belief in the dynamic approach to belief revision: an agent believes that a proposition P is true at the current state, iff P is true at the most plausible state with respect to the plausibility ordering for the agent at the current state. Hierarchies of belief are interpreted by hierarchies of plausibility order in the model. This approach is a *qualitative* approach to belief. Degrees of beliefs are studied by a *quantitative* approach in [33], for instance. In game theory, hierarchies of beliefs are introduced for an epistemic analysis of solution concepts in dynamic games (e.g., [2]).

Our approach to the abductive search for plausible conjectures suggests that some hierarchies or orderings of plausible conjectures can indeed be introduced between conjectures. But the creation of those orderings is not to be taken in any belief-revisionary fashion due to their deflationary, Bayesian, or even the IBE-related flavours. If one were to ground such ranks on probabilistic methods, many novel guesses (for which the misnomer of 'high-risk-high-gain' is in these days often used) that fall outside of standard deviations would be weeded out. This would merely serve to block the growth of inquiry. To aid narrowing down very large spaces of plausible conjectures and to so eliminate uncertainty concerning the myriad of outcomes that abductive reasoning is prone to produce, one could instead conceive of various plausibility measures and orderings between conjectures with such weighings. In those cases the more plausible hypotheses could be taken to be those that survive ! longer, or more precisely speaking are the ones that survive even when faced with a larger number of or more weighty surprising facts than many other hypotheses do. Such plausible hypotheses are those that scientists continue to uphold by putting their hopes on ones that are anticipated to turn out to be the case in the future, albeit perhaps only in the distant posteriority. But the crucial point is that the increase in plausibility need not imply any increase in the inquirer's belief in them.[14] The ranking between plausible conjectures would follow from the proportion of hypotheses kept on board to those that are dropped, given the amount of different but connected surprising facts encountered during the related abductive reasoning processes, plus the all-important economic and strategic factors relevant to the proportional selection to test those hypotheses.[15] No strong epistemic element is involved here: we do not work with knowledge or belief concerning hypotheses and conjectures. What is crucial is how plausibility relates not to any single guess but to the network of related, connected guesses and their anticipated outcomes. To increase plausibility o! f one's guesses one could for instance look for possible connections between the outcomes of a number of other, related guesses in a number of other, related abductive tasks. We leave deeper analyses of these issues for further work.

9 Conclusion

In his late theory, Peirce called abduction "reasoning from surprise to inquiry". In his unpublished 1905 letter he attributes a novel logical form to it which is that of the Modus Tollens and in which the conclusion is put in the interrogative ("investigand") mood. The conclusion does not present a given hypothesis for contemplation; it merely suggests that it would be reasonable and prudent, given the overall goals, background theories and assumptions, common experience and the context of investigation, to inquire whether a suggested hypothesis could turn out to be the case or not. Peirce's novel schema proposes adoption of a hypothesis as a reasonable thing to do in the sense of promising specific strategic advantages to those who follow their intellectual courage to guess at that hypothesis.

In this paper, we provided a new interpretation of Peirce's insights concerning the logical form of abduction, and developed those insights further as detailed in the dynamic logic of support. That logic incorporates the indispensable dynamic

---

[14]Poignant remarks from Peirce attest this: "Belief has no place in science at all ... Nothing is *vital* for science; nothing can be. Its accepted propositions, therefore, are but opinions at most; and the whole list is provisional. The scientific man is not in the least wedded to his conclusions. He risks nothing upon them. He stands ready to abandon one or all as soon as experience opposes them" (CP 1.635).

[15]Fear no failures, since the cost of questions is small: "In pure abduction, it can never be justifiable to accept the hypothesis otherwise than as an interrogation. But as long as that condition is observed, no positive falsity is to be feared; and therefore the whole question of what one out of a number of possible hypotheses ought to be entertained becomes purely a question of economy" (CP 6.528, c.1901; cf. [19]).

aspects of abductive reasoning and analyzes what the conjecture-making in Peirce's sense could logically speaking mean. After all, the goal of Peirce's 1905 abductive schema is not to abduce any singular proposition to serve as an explanation of facts. Its aim is to conclude which overall and general lines of investigation and research agendas would be the most reasonable ones for scientists to adopt. In that very concrete sense, abduction has a much wider reach than what has ordinarily been recognized. It suggests some preliminary pointers and directions towards what could evolve into a complicated process of interlocking procedures for theory formation. It does not give closure on the sets of propositions that one is to believe in at the expense of some others.

References

[1] A. Baltag and S. Smets. A qualitative theory of dynamic interactive belief revision. In G. Bonanno, W. van der Hoek and M. Wooldridge, editors, *Logic and the Foundations of Game and Decision Theory* (LOFT 7). pp.11–58. Amsterdam University Press, 2008.

[2] P. Battigalli and M. Siniscalchi. Hierarchies of conditional beliefs and interactive epistemology in dynamic games. *Journal of Economic Theory*. 88(1):188–230, 1999.

[3] F. Bellucci and A.-V. Pietarinen. The Iconic Moment. Towards a Peircean theory of diagrammatic imagination, to appear in O. Pombo, A. Nepomuceno and J. Redmond (eds.), *Epistemology, Knowledge, and the Impact of Interaction*, Springer, Dordrecht, pages 463–481, 2016.

[4] J. van Benthem and M. Martinez. The stories of logic and information. In P. Adriaans and J. van Benthem, editors, *Philosophy of Information*, Handbook of the Philosophy of Science, pages 217–280. 2008.

[5] D. Campos. On the distinction between Peirce's abduction and Lipton's Inference to the best explanation. *Synthese*. 180:419–442, 2011.

[6] B. Chellas. *Modal Logic: an introduction*. Cambridge University Press, 1980.

[7] I. Ciardelli, J. Groenendijk, and F. Roelofsen. Inquisitive semantics: a new notion of meaning. *Language and Linguistics Compass*. 7:459-476, 2013.

[8] J. Dewey. *How We Think*. D.C. Heath & Co., Boston, 1910.

[9] D. M. Gabbay & J. Woods. *The Reach of Abduction. Insight and Trial*. Elsevier, Amsterdam.

[10] V. Goranko and S. Passy. Using the universal modality: gains and questions. *Journal of Logic and Computation*. 2(1):5–30, 1992.

[11] N. R. Hanson. *Patterns of Discovery*. Cambridge University Press, Cambridge, 1958.

[12] Y. Hamami. The interrogative model of inquiry meets dynamic epistemic logics. *Synthese*. 192:1609–1642, 2015.

[13] C. S. Hardwick and J. Cook (editors). *Semiotic and Significs: The correspondence between Charles S. Peirce and Victoria Lady Welby*. Indiana University Press, Blooming-

ton, 1977.

[14] J. Hintikka. *Socratic Epistemology. Explorations of knowledge-seeking by questioning.* Cambridge University Press, Cambridge, 2007.

[15] M. Kiikeri. Abduction, IBE and the discovery of Kepler's ellipse. Explanatory Connections. In P. Ylikoski and M. Kiikeri, editors, *Electronic Essays Dedicated to Matti Sintonen*, 2001. http://www.helsinki.fi/teoreettinenfilosofia/henkilosto/Sintonen/Explanatory%20Essays/kiikeri.pdf

[16] S. A. Kleiner. A new look at Kepler and abductive argument. *Studies in History and Philosophy of Science.* 14:279–313, 1983.

[17] M. Ma and K. Sano. How to update neighborhood models. In D. Grossi, O. Roy, and H. Huang, editors, *LORI 2013, volume LNCS 8196*, pages 204–217, Springer, Berlin & Heidelberg, 2013.

[18] I. Niiniluoto. Abduction, tomography, and other inverse problems. *Studies in History and Philosophy of Science.* Part A, 42(1):135–139, 2011.

[19] S. Paavola. Abduction as a logic and methodology of discoveries: The importance of strategies. *Foundations of Science.* 9(3): 267–283, 2004.

[20] S. Paavola. Hansonian and Harmanian abduction as models of discovery. *International Studies in the Philosophy of Science.* 20:93–108.

[21] C. S. Peirce. Deduction, Induction, and Hypothesis. *Popular Science Monthly.* August 1878, pages 470–482. Reprinted in [25, pp. 323–338], Volume 3.

[22] C. S. Peirce. A Theory of Probable Inference. In *Studies in Logic by Members of Johns Hopkins University*, ed. by C. S. Peirce, 1883, pages 126–181, Little & Brown, Boston, MA.

[23] C. S. Peirce. *The Collected Papers of Charles S. Peirce*, 8 vols., ed. by Hartshorne, C., Weiss, P. and A. W. Burks. Cambridge: Harvard University Press, 1931–1966. Cited as CP followed by volume and paragraph number.

[24] C. S. Peirce. Manuscripts and Letters in the Houghton Library of Harvard University, as identified by Richard Robin, Annotated catalogue of the papers of Charles S. Peirce, University of Massachusetts Press, Amherst, 1967, and in The Peirce Papers: A supplementary catalogue, *Transactions of the C. S. Peirce Society.* 7:37–57, 1971. Cited as MS or MS L, followed by manuscript number and, when available, page number.

[25] C. S. Peirce. *Writings of Charles S. Peirce: A Chronological Edition*, 7 vols., ed. by E. C. Moore and C. J. W. Kloesel et al. Bloomington: Indiana University Press. Cited as W followed by volume and page number.

[26] A.-V. Pietarinen. Guessing at the unknown unknowns. In T. Nakajima and T. Baba, editors, *Contemporary Philosophy in the Age of Globalization: Hawai'i Conference*, Vol. 3, Institute for Advanced Study on Asia, Tokyo, pages 87–100, 2014.

[27] A.-V. Pietarinen. The science to save us from philosophy of science. *Axiomathes*, 25:149–166. DOI 10.1007/s10516-014-9261-8

[28] A.-V. Pietarinen (editor). *Logic of the Future. Peirce's writings on existential graphs.* Indiana University Press, Bloomington, to appear.

[29] A.-V. Pietarinen and G. Sandu. IF Logic, Game-theoretical Semantics, and Philosophy of Science. S. Rahman, D. Gabbay, J. P. Van Bendegem, J. Symons (eds), *Logic, Epistemology and the Unity of Science*. Dordrecht: Kluwer, 105–138, 2004.

[30] A.-V. Pietarinen and F. Bellucci. New light on Peirce's conceptions of retroduction, deduction, and scientific reasoning. *International Studies in the Philosohy of Science*. 28(4):1–21, 2014. http://dx.doi.org/10.1080/02698595.2014.979667

[31] S. Psillos. An explorer upon untrodden ground: Peirce on abduction. In *Handbook of the History of Logic. Volume 10: Inductive Logic*. Ed. by D. M. Gabbay, S. Hartmann, and J. Woods. Elsevier, Oxford, pages 117–151.

[32] J. van Benthem and S. Minică. Toward a dynamic logic of questions. *Journal of Philosophical Logic*. 41(4):633–669, 2012.

[33] H. P. van Ditmarsch. Prolegomena to dynamic logic for belief revision. *Synthese*. 147(2):229–275, 2005.

[34] F.R. Velázquez-Quesada, F. Soler-Toscano and Á. Nepomuceno-Fernández. An epistemic and dynamic approach to abductive reasoning: Abductive problem and abductive solution. *Journal of Applied Logic*. 11:505–522, 2013.

[35] C. Wilson. Newton and some philosophers on Kepler's laws. *Journal of the History of Ideas*. 35:231–258, 1974.

[36] J. Woods. Recent developments in abductive logic. *Studies in History and Philosophy of Science, Part A*. 42(1):240–244, 2011.

Received January 2015

Abduction and Beyond
Methodological and Computational
Aspects of Creativity in Natural Sciences

Andrés Rivadulla*

Complutense University Madrid, Spain
arivadulla@filos.ucm.es

Abstract

My first aim in this paper is to complement *abduction* as a standard discovery practice in the methodology of science. I do this in two ways: firstly by means of the introduction of *preduction* as a deductive discovery strategy. I term *preduction* the method of reasoning, used predominantly in theoretical physics, which starting from the available theoretical background as a whole, allows for the deductive *anticipation* of previously unknown results, provided that the mathematical combination of already accepted results of different theories or disciplines is compatible with dimensional analysis. This is the method by which many hypotheses, laws, and theoretical models are introduced into physics. *Preduction* complements Charles Peirce's view that it is only via abduction that new ideas come to science. But, secondly, since standard abduction is not always sufficient to provide best explanations, the methodology of science sometimes has to proceed *more preductivo*: abduction resorts to preduction. I call this form of abductive reasoning *sophisticated abduction*.

Moreover, the automation of ampliative inferences such as induction and abduction in the field of computational discovery of scientific knowledge encourages the hypothesis of *computational preduction*. *Computational preduction* extends the possibilities of machine learning, which in the past forty years has facilitated the possibilities of computational systems both to rediscover empirical laws and to automatically discover equations in data bases. Automated preduction should also facilitate scientific creativity in theoretical physics.

Keywords: Scientific Discovery, Standard Abduction, Theoretical Preduction, Sophisticated Abduction, Automated Discovery, Computational Preduction

I very much thank the four anonymous referees for their interesting comments on an earlier version of this paper.

*Complutense research Group 930174 and research project FFI2014-52224-P financed by the Ministry of Economy and Competitiveness of the Kingdom of Spain.

1   Introduction

Throughout the twentieth century, philosophers of science have to a great extent abandoned the issues relating to the context of scientific discovery. Popper's negation of the methodological relevance of the ways by which new ideas enter into science has contributed to ignorance of the context of discovery in the methodology of science. Most philosophers have consequently disregarded the creative role played in natural science by ampliative inferences such as induction and abduction. Moreover, the consolidation of the hypothetic-deductive method in the methodology of science in the twentieth century has ironically contributed to the neglect of the significance of deductive reasoning in the context of discovery. As Peter Medawar acknowledged in [14], his contribution to Schilpp's volume dedicated to Karl Popper, "disclaiming any power to explain how hypotheses come into being" was a sign of the weakness of hypothetic-deductivism.

In spite of this negative attitude, abduction has played a fundamental role in the history of Western science for the introduction of highly significant hypotheses, both in observational and in theoretical natural sciences, from the postulation of models of the planetary movements in ancient astronomy to the most recent hypotheses on dark matter and dark energy in theoretical astrophysics. Section 2 is intended to show the importance of standard abduction in the methodology of natural sciences.

In Section 3 I argue that deductive reasoning can be extended to the context of discovery of theoretical natural sciences such as mathematical physics. I use the term *theoretical preduction* to describe the method of reasoning which consists in the implementation of deductive reasoning in the context of scientific discovery. My aim will be to show the role that preduction plays in the methodology of physical sciences. I claim that preduction is a method of reasoning which, starting from the available theoretical background as a whole, allows for the deductive anticipation of new results, provided that the combination and mathematical manipulation of the previously accepted results of different disciplines of theoretical physics - those which have been taken as premises of preductive reasoning - are compatible with dimensional analysis. I affirm that this is the way in which many factual hypotheses, laws, and theoretical models are postulated in physics. Since the results, which are the premises of preductive reasoning, derive from different theories, preduction represents transverse or inter-theoretical deduction. This is what makes it possible to anticipate new ideas in physics.

In Section 4 I tackle the role of abduction in the context of explanation. It is widely known that during the 1960s, abduction was identified as a form of inference to the best explanation. Nonetheless, the explanation of many natural phenomena requires more than spontaneous acts of creativity such as those that one might imagine in the cases of Kepler, Darwin or Wegener. Much mathematical work often has to be done in order to advance justified theoretical explanations. In those cases, the explanation takes place *more preductivo*. I term *sophisticated abduction* the corresponding inference to the best explanation, and I illustrate this fact of the methodology of science by resorting to an interesting case study of theoretical astrophysics: stellar pulsation.

Finally in Section 5 I address the question of discovery from the point of view of machine discovery theory. Since the 1970s, despite the neglect of scientific discovery by most official philosophers of science, the field of computer-supported scientific discovery has attracted the attention of scientists working in the domain of artificial intelligence and knowledge acquisition, meaning that the field of computational discovery of scientific knowledge has become increasingly important and fruitful. Indeed, according to [30, 1]: "Whereas early work in the philosophical tradition emphasized the *evaluation* of laws and theories (e.g., Popper 1965 [*Conjectures and Refutations*]), recent research in the paradigm of cognitive science has emphasized scientific *discovery*, including the activities of theory formation, law induction, and experimentation. Moreover, the early philosophical approaches focused on the *structure* of scientific knowledge, whereas recent work has focused on the *process* of scientific thought and on describing these activities in *computational* terms".

Once ampliative inferences such as induction and abduction have been implemented in the field of machine learning, the next step will encourage the computational hypothesis of theoretical preduction. Computational preduction equates to automating preduction mechanisms. This possibility complements any already existing computational methods for the automation of scientific discoveries. Automated preduction would facilitate the anticipation of new results in theoretical physics. The philosophy of science invites discovery science to start with the possibility of automated preductive discovery in theoretical sciences such as mathematical physics.

A complete analysis of abduction – abductive reasoning has many faces, and is a very polyhedric instrument – should make reference to the underlying logic of abduction – an enterprise which is being tackled by many contemporary logicians in a very competent way– as well as to the use of abduction at the meta-methodological level. Indeed scientific realists argue nowadays that realism is grounded in a second order or grand abduction. But since both these questions constitute separate domains of contemporary philosophy, they fall outside the scope of this article.

2   Standard abduction.   The methodology of discovery in natural sciences

Is the question of how it happens that a new idea is introduced into science relevant to the methodology of science? There are two different answers to this question:

1. The negative one, the *Popper-Reichenbach approach*, according to which the philosophy of science is concerned only with questions of justification or validity, i.e. only how scientists test their theories, hypotheses or conjectures is methodologically relevant. How they find them is irrelevant.

2. The positive one, the *Peircean approach*

Following Popper-Reichenbach's view, the contemporary official methodology of science has focused exclusively on synchronic or systematic aspects of science (the justification context), and has disregarded for decades the issues relating to the methods of conceiving new ideas (the discovery context). According to the official view, science applies exclusively the deductive testing of hypotheses (Popperian tests).

Nonetheless, in some natural sciences, for instance in geology, we find the following viewpoints. Dan McKenzie, one of the creators of the plate tectonics model, confesses in [13, p.185] that: "hypothesis testing in its strict form is not an activity familiar to most earth scientists", and John Sclater [28, p.138] affirms: "Earth scientists, in most cases, observe and describe phenomena rather than conducting experiments to test hypotheses". And in palaeoanthropology, James Noonan, Pääbo *et al.* [16] in a paper on the comparison of the genomes of Neanderthals and modern humans, claim that "Our knowledge of Neanderthals is based on a limited number of remains and artefacts from which we must make inferences about their biology, behavior, and relationship to ourselves". For his part, Carlos Lorenzo [10, p.103] affirms that "Philogenetic trees are only evolutionary hypotheses built upon a continuously changing empirical basis. It is usual that these hypotheses are tested, and modified, if necessary, on the ground of new data". Thus deductive testing is not the exclusive methodology in natural sciences. It may not even be the predominant one.

Thirty years before Popper and Reichenbach's neglect of the philosophical relevance of the processes of conceiving or inventing scientific hypotheses, Charles Peirce took up a position on the processes of forming or devising explanatory scientific hypotheses. Peirce's basic idea was very simple: more often than not, scientists stumble over unexpected, surprising, striking facts or phenomena. Instead of passively wondering about them, it seemed reasonable to Peirce to adopt an approach both logical and philosophical about how to propose or to introduce new hypotheses: an approach which, if true, would account for the facts. Peirce gave the name *abduction* to the logical operation which introduced new ideas into science. For instance: "All the ideas of science come to it by the way of abduction. Abduction consists in studying facts and devising a theory to explain them" [18, 5.170]

Popper-Reichenbach's negation of the importance of the context of discovery was one of the major errors of contemporary philosophy of science. Indeed, abduction has been widely applied in the natural sciences, from ancient astronomy to contemporary theoretical astrophysics, evolutionary biology, the earth sciences and palaeontology, among others. Let me illustrate some of them in more detail.

From a methodological viewpoint, palaeoanthropology behaves like a typical empirical science: recognition of surprising facts, abduction by elimination of mutually exclusive hypotheses, hypothesis revision in the light of novel data, and the beginning of a new cycle. For instance, the postulation of a new hominin species, *Homo antecessor*:

1. The surprising fact was the discovery on 1995 at Sierra de Atapuerca, Spain, of a part of the facial skeleton of a young man – the *Gran Dolina Boy* – with an antiquity of nearly 800 kiloyears.

2. This fossil did not show either the primitive features of *Homo ergaster* or the derived characters of *Homo heidelbergensis* – 500 kiloyears old – which were inherited by *Homo neanderthalensis*.

3. Moreover, the skull capacity of the *Gran Dolina Boy* – about 1000 cc, considerably bigger than that of the best preserved skulls of *Homo ergaster* – provided an excellent reason for not considering it to be a member of *Homo ergaster*.

4. Furthermore, the *Gran Dolina Boy* could not be a representative of *Homo erectus*, since this is mainly distributed throughout Asia (also in Israel and in Georgia), and these fossils are considerably older.

5. Conclusion: The *Gran Dolina Boy* could only belong to a new species, *Homo antecessor*: "a common predecessor of both the evolutionary line that in Europe led to *Homo neanderthalensis* and of the evolutionary line that in Africa led to the modern populations of *Homo sapiens*". [2, p.35]

As for the case of the *continental drift* hypothesis in the earth sciences, the postulation of this hypothesis has proceeded – admits Alfred Lothar Wegener [34, p.167] – purely empirically: "by means of the totality of geodetic, geophysical, geological, biological and palaeoclimatic data. ... This is the *inductive method*, one which the natural sciences are forced to employ in the vast majority of cases". (Both the name and the logic and philosophy of *abduction* were completely unknown in 1915, the year of the first edition of *Die Entstehung der Kontinente und Ozeane*). A huge number of facts which suggested this hypothesis to Wegener is summarily presented here (Cf. also [25]):

1. *Geodetic data*: Observation, on the basis of astronomical, radiotelegraphic and radio-emission measures, of the continuous separation of Europe and America.

2. *Geophysical data*: Compatibility of the *Fennoscandian rebound* and the *isostasy* hypothesis with lateral continental displacements.

3. *Geological data*: Affinities between the plateaus of Brazil and Africa, and between the mountains of Buenos Aires and the Cape region, etc.

4. *Palaeontological data*: The distribution of the Glossopteris flora – a fern fossil register – in Australia, South India, Central Africa and Patagonia, and of Mesosaurus in Africa and South America

5. *Palaeoclimatical data*: The Spitzberg(en) Islands, nowadays affected by a polar climate, must have enjoyed a much warmer climate in the Mesozoic and in the Palaeozoic.

As these examples show, observational natural sciences like palaeoanthropology and the Earth sciences are predominantly empirical. They rely on experience, on observation, they apply abduction as a discovery practice. They do not implement any Popperian deductive testing of hypotheses but only, when required, additional evidence tests.

Nonetheless, as the following example shows, abduction is also applied in theoretical sciences as the only form of introducing new ideas into science. In astrophysics, the Newtonian version of Kepler's Third Law is a very useful instrument. It permits, for instance, the calculation of the Sun's mass and the masses of the planets of our solar system; it allows the calculation of the masses of the stars that constitute a binary system; and last but not least, it allows the determination of the number of stars of our Galaxy. It is a very successful law indeed. Moreover, Kepler's Third Law demands that any heavenly body situated far away from its rotation centre move very slowly. Thus, according to the law, stars located far away from their galaxy centres must have very small orbital velocities.

In the 1930s, Jan Hendrik Oort and Fritz Zwicky discovered to their surprise that this was not the case. Instead of declaring Kepler's Third Law to be wrong

– and therefore deducing that Newtonian mechanics is mistaken – Zwicky, in an undoubtedly abductive manner, proposed that there must be inside galaxies much more matter than can be observed. Since he had no idea how this hidden matter was constituted, he coined for it the now-famous term *dark matter*. As we will see according to the nomenclature to be introduced in Section 4, the hypothesis of the existence of *dark matter* was the best explanation of the (surprising) fact that stars located on the boundaries of their galaxies have an orbital velocity that is higher than expected. I illustrate further abductive inferences in theoretical cosmology in [26].

3 The boundaries of abduction. Preductive practices of creativity in theoretical natural sciences

For the sake of argument I return to Charles Peirce [18, 5.145]: "Induction can never originate any idea whatever. No more can deduction". Nonetheless I do not completely agree with him that "All the ideas of science come to it by the way of abduction" [18, 5.145]. Thus my main question is: *can deductive reasoning be used in the context of scientific discovery?* My answer will be: Yes, it can.

I maintain that in the methodology of theoretical physics, we can implement deductive reasoning in the context of discovery, beyond its ordinary uses in the context of justification. This is possible because theoretical physics uses mathematics as an indispensable tool. Theoretical physicists apply at will Leibniz's principle of *substitutio salva veritate*: dimensional analysis guarantees *substitutio salva legalitate*, i.e. the legitimacy of the undertaken substitutions. Thus a new form of reasoning in scientific methodology, which I call *theoretical preduction* or simply *preduction*, can be identified.

Peter Medawar [14, p.289] pointed out the weakness of deductive reasoning in scientific methodology: "The weakness of the hypothetico-deductive system, insofar as it might profess to offer a complete account of the scientific process, lies in its disclaiming any power to explain how hypotheses come into being". Later, Thomas Nickles [15, p.446] conjectured about the applicability of deductive reasoning in scientific discovery: "it is worth noting that even an ordinary deductive argument need not be sterile: it may be epistemologically ampliative even though it is not logically ampliative, for we are not logically omniscient beings who see all the logical consequences of a set of propositions". But it was Elie Zahar [35, 244-245] who unambiguously advanced the thesis of the deductive character of scientific discovery: "*the process of discovery ... rests largely on deductive arguments from principles which underlie not only science and deductive metaphysics but also everyday de-*

cisions" (my italics). On pages 249-250 Zahar further argues that "the logic of scientific discovery is not inductive; neither does it resemble artistic creation, but is, instead, largely deductive. ... Unlike Popper, I think that deduction constitutes the most important element in the process of invention". But since Zahar's thesis of the deductive character of scientific discovery is based on the metaphysical assumption of several *philosophical principles* and *meta-principles*, which, as a kind of premise for deductive reasoning, allow for the creation of theories – the *Principle of Correspondence,* the *Principle of Identity,* the *Principle of Sufficient Reason,* and the *Principle of Proportionality of Cause and Effect* – his proposal becomes a metaphysical theory about the deductive nature of scientific discovery which cannot be made congruent with both ampliative and anticipative strategies in the methodology of natural sciences.

1. *Preduction,* is a method of reasoning that begins with previously accepted results of the available theoretical background. It consists in resorting to the available results of theoretical physics as a whole, in order to *anticipate* new ideas by mathematical combination and manipulation, compatible with dimensional analysis, of the available results, although not every combination need be heuristically fruitful.

2. The results postulated *methodologically* as premises of an inferential procedure proceed from differing theories, and any accepted result can serve as a premise – on the understanding that *accepted* does not imply *accepted as true.*

3. This suggests the notion of a hypothetic-deductive method. Indeed, *preduction* is an implementation of the deductive way of reasoning. The specificity of preduction is that it is an extension of deductive reasoning into the contexts of scientific discovery (and explanation).

4. Since the results, which are the premises of the preductive way of reasoning, proceed from different theories, preduction is *transversal* or *inter-theoretical* deduction. This is what makes it possible to anticipate new ideas in physics, i.e. deductive creativity in science.

5. Preductive reasoning differs from standard abduction, and from any ampliative inference, in that the preduced results do not proceed from empirical data, but rather follow from deductive reasoning starting from the available accepted theoretical background taken as a whole.

Abduction is certainly a very important form of reasoning to the methodology of science. But it is not enough, for new ideas: hypotheses, laws, theoretical models, etc. may also be advanced deductively. So abduction must be complemented by preduction as forms of scientific discovery. Abduction and preducción are complementary strategies of scientific creativity. Other characterizations of the concept of theoretical preduction, together with illustrative examples from physics, I have presented in Rivadulla [22, 24, 23, 25].

Physicists apply preductive reasoning in a natural way in order to anticipate as yet unavailable ideas. Stellar astrophysics can be considered paradigmatic for the application of preductive reasoning. The British physicist Arthur S. Eddington [3, 1] affirms, for instance, that although "At first sight it would seem that the deep interior of the sun and stars is less accessible to scientific investigation than any other region of the universe..., the interior of a star is not wholly off from such communication. A gravitational field emanates from it ...; further, radiant energy from the hot interior after many deflections and transformations manages to struggle to the surface and begin its journey across space. From these two clues alone *a chain of deduction can start, which is perhaps the more trustworthy because it is only possible to employ in it the most universal rules of nature –the conservation of energy and momentum, the laws of chance and averages, the second law of thermodynamics, the fundamental properties of the atom, and so on.* There is no more essential uncertainty in the knowledge so reached than there is in most scientific inferences" (my italics). And Dale A. Ostlie & Bradley W. Carroll [17, p.590] claim, with regard to the Chandrasekhar limit formula for the mass of white dwarfs: "This formula is truly remarkable. It contains three fundamental constants $-\hbar$, c, and G – representing the *combined effects of quantum mechanics, relativity, and Newtonian gravitation* on the structure of a white dwarf" (my italics). These quotations confirm that, if we wish to postulate theoretical models in stellar astrophysics, we have to combine available relevant results from different disciplines of theoretical physics, so as to derive an equation, or a set of coupled equations, relating to the phenomenon under investigation.

As a matter of fact the *discovery* of the interior structure of main sequence stars amounts to producing a theoretical model of the stellar interiors. This model consists of five basic differential equations: pressure, mass, luminosity and temperature (twofold) gradients (Cfr. [17, ch.10]). From a methodological viewpoint the most interesting equations of this theoretical model are the equations of hydrostatic equilibrium and of temperature gradient.

The idealisations needed for hydrostatic equilibrium are that of a spherically symmetric and static star. The corresponding *preductive* procedure consists in the *combination of three theories*: Newtonian mechanics (second and third laws, and the universal gravitation law), classical statistics mechanics (the Maxwell-Boltzmann distribution of the ideal gas pressure), and quantum physics (Planck's radiation law of the radiative pressure of a black body).

The idealisations, assumed for the obtaining of the temperature gradient, are also that of a static sphere with black body conditions plus the conditions of adiabatic expansion. The corresponding *preductive* procedure consists in the *combination of three disciplines*: classical physics for the temperature gradient of radiative transfer (combination of the equation of radiative transfer with the equation of the black body radiation pressure); classical statistical mechanics; and thermodynamics of adiabatic processes (for the obtaining of the temperature gradient of a gas expanding adiabatically).

A full account of the theoretical model of stellar interiors should be completed with the mass conservation equation and the equation of the luminosity gradient, with the latter depending on the energy generated both by nuclear and gravitational processes.

4 Getting across the frontiers. Sophisticated abduction

Abduction has to two sides. I completely agree with Lorenzo Magnani [12, p.294] that abduction both generates plausible hypotheses and that successful abduction provides best explanations of facts. Magnani [11, p.17-18] anticipates this idea as he recognizes that "Theoretical abduction is the process of *inferring* certain facts and/or laws and hypotheses that render some sentences plausible, that *explain* or *discover* some (eventually new) phenomenon or observation; it is the process of reasoning in which explanatory hypotheses are formed and evaluated".

Indeed according to Peirce [18, 7.202] abduction, as "the step of adopting a hypothesis as being suggested by the facts", also supplies an explanation for novel or surprising phenomena: "The explanation must be such a proposition as would lead to the prediction of the observed facts, either as necessary consequences or at least as very probable under the circumstances. A hypothesis then, has to be adopted which is likely in itself and renders the facts likely". Gilbert Harman [7, p.88-89] and [8, p.165] prefers to name abduction *inference to the best explanation* to point to the fact that abduced hypotheses provide the best explanation for the evidence than would any else.

Since the late 1970s, abduction has been, for the methodology of science, inference to the best explanation. Paul Thagard [33, p.77] claims that "Inference to scientific hypotheses on the basis of what they explain was discussed by such nineteenth-century thinkers as William Whewell and C. S. Peirce ... To put it briefly, inference to the best explanation consists in accepting a hypothesis on the grounds that it provides a better explanation of the evidence than is provided by alternative hypotheses". Thagard [33, p.77-78] illustrates this viewpoint resorting to Fresnel's wave theory of light, which "explained the facts of reflection and refraction at least as well as did the particle theory, and ... there were other facts,..., which only the wave theory could simply explain. (...) Hence the wave theory should be inferred as the best explanation". And Hilary Putnam [19, p.198] confesses that "we accept the Darwinian theory of evolution by natural selection as what Peirce called an 'abduction' or what has recently been called an 'inference to the best explanation'". In the contemporary theory of abduction John Josephson [9, 5] for instance claims that "*Abduction, or inference to the best explanation, is a form of in-ference* that goes from data describing something to a hypothesis that best explains or accounts for the data".

In Section 2 I have focused mainly on the creative aspect of abduction. As a form of scientific discovery, abduction has made it possible to introduce very significant hypotheses in science during the last two and a half millennia, hypotheses that also fit best to – i.e. best explain – the facts. Nonetheless sometimes creative/explanatory acts are not as simple as it might seem.

Creativity in science is not always spontaneous. If for the sake of illustration we restrict ourselves to theoretical physics, then very frequently hard physical-mathematical work is needed when the available empirical data do not directly suggest an attractive explanation. Since moreover theoretical physicists nowadays generally combine theoretical and empirical results accepted in differing branches of physics, preductive reasoning unavoidably becomes subsidiary to abduction when providing satisfactory explanations for as yet unexplained constructs. This is why I say (cf. [24, p.361]) that in such cases the theoretical explanation takes place *more preductivo*. Since in such cases the inference to the best explanation depends on the implementation of preductive reasoning on the context of theoretical explanation, I name this procedure *sophisticated abduction*.

In order to understand this proposal as unambiguously as possible, it is important to note that explanation in theoretical physics encompasses more cases than are sup-posed to be subsumed under Popper-Hempel's law-covering nomological-deductive model of scientific explanation. As I claim in [21, p.166-167] "any physical construct: facts, laws, hypotheses, etc., can only receive a theoretical explanation, when it can be deduced mathematically in the framework of another physical construct of higher level. Thus not only facts, but also empirical generalizations, abstract laws and even theories themselves admit of explanations in this sense". For instance, Kepler's em-pirical laws of planetary movements was theoretically explained only after it was incorporated into Newtonian celestial mechanics; Stefan's Law of black body radia-tion and Wilhelm Wien's displacement law received a theoretical explanation when they could be deduced mathematically from Max Planck's radiation law; and Bohr's atomic model for the first time provided a theoretical explanation of the spectra of elements etc. But even the principles of phenomenological thermodynamics could be satisfactorily explained, as they were justified within the framework of statistical thermodynamics. This concept of theoretical explanation realizes Einstein's ideal of hypothetic-deductive physics as he develops it in different articles reprinted in Einstein's [4].

Since abduction belongs not only to the context of discovery but also to the context of scientific explanation – they are the twin faces of abduction – explanatory abduction is committed to the norms of theoretical explanations. In [24, §4] I have given two examples of sophisticated abduction in the realm of quantum mechanics. In the following I am going to present, very summarily, a different example taken from the recent history of theoretical astronomy.

At the end of the sixteenth century, between 1595 and 1596, some fifteen years before Galileo's astronomical observations, the German astronomer David Fabricius (1564-1617) observed a variation in the brightness of star o Ceti in the constellation Cetus. Had he been during his lifetime a better known astronomer, this discovery would have contributed to the refutation of the Aristotelian theory of the immutability of heavens. The star o Ceti was named *Mira, admirable* in Latin, by the Polish astronomer Johannes Hevelius (1611-1687). Finally, in 1667 the French astronomer Ismael Boulliau (1605-1694) determined the period of eleven months of *Mira Ceti*. Later, in 1783, John Goodricke (1764-1786) discovered the variability of δ *Cephei*, and Edward Pigott (1753-1825) in 1784 the variability of η *Aquilae*. These variable stars came to be known as *Cepheids*. At the end of the nineteenth century, Henrietta Swan Levitt (1868-1921) discovered more than two thousand variable stars, most of them in the Magellanic Cloud. But the phenomenon of the variability of *Cepheids* demanded an explanation.

Several hypothetical explanations were suggested. The first ones made use of an analogy argument: the variability could be due to the existence of spots like sunspots. Later, Goodricke, Belopolsky and even Schwarzschild proposed that the variability was due to the binary character of the stars. The tidal effects in the atmospheres of the stars might be responsible for the variable brightness observations. But in his *The Internal Constitution of the Stars*, 1926, Arthur Eddington stated that the American astronomer Harlow Shapley (1885-1972) had put forward, in [29], "the chief arguments against the binary theory and in favour of the pulsation theory". According to [3, 184], "the binary model is a geometrical impossibility". The reason was that, according to the already accepted astrophysical background, the radii of the binary stars should be larger than the radii of their own orbits. Three years after Shapley's proposal Eddington himself – in two articles published in *Observatory* 40, 1917 and in *Monthly Notices of the Royal Astronomical Society* 79, 1918 – provided a solid theoretical argument on behalf of the pulsation hypothesis. Henceforth variable stars became pulsating stars. The issue is that, from a methodological viewpoint, the explanation of stellar pulsation cannot be conceived in terms of standard abduction. Much more than a creative jump is involved here. On my view the explanation of stellar pulsation is a typical process of sophisticated abduction, since it requires the preductive construction of a theoretical model for a pulsating variable star, and this is only possible when suitable results from different branches of theoretical physics are taken into account and mathematically combined with one another. This procedure is by no means simple.

The first stimuli came from a paper, already old at the time, by the German astrophysicist August Ritter, "Investigations on the height of the atmosphere and on the constitution of gaseous heavenly bodies"[20]. In this article, equation (245), Ritter claims that the pulsation period of a supposed uniformly dense gaseous sphere is inversely related to the square root of its density. Modern theoretical models of stellar pulsation start with Eddington's suggestion in [3, p.186] that, for reasons of simplicity, "the oscillations through the greater part of the interior are approximately adiabatic. We therefore start by considering adiabatic oscillations of a sphere of gas". The resulting model is, like in Ritter's case, an equation expressing that the pulsation period of a variable star is inversely related to the square root of its average density. In order to achieve this result, it is necessary to combine (Cf. [17, §14.2] or [1, §16.4] classic thermodynamics (first law) with acoustics and wave theory (adiabatic velocity of sound) and classical mechanics (hydrostatic equilibrium). According to Ostlie & Carroll [17], the resulting equation provides a quantitative agreement with the observed periods of Cepheids that is "not too bad", and for Erika Böhm-Vitense [1], this equation "explains the observed period-luminosity relation". The bigger the star, the lower its density, and since the period is inversely related to the density, the star's oscillation will be quicker. This is a typical phenomenon with big stars.

This explanation is the best explanation astrophysicists can admit for stellar pulsation. It is an abductive one. But since it proceeds in a preductive way, it is not standard abduction, it is sophisticated abduction. Preduction lies at the other side of the border line from abduction. Sophisticated abduction, which uses preduction, is an excellent example of the collaboration of both kinds of reasoning, abduction and preduction, in scientific creativity.

5   Computational models of scientific discovery

The existence of computational studies of science makes evident the lack of imagination of many academic philosophers of science, including both the pro-positivist

Hans Reichenbach and the anti-positivist Karl Popper, who neglected the importance of the *context of scientific discovery* in favour of the *context of justification*. The ignoring of scientific discovery left the field open for researchers in cognitive science and artificial intelligence to develop the computational study of science. Indeed, Shrager and Langley [30, 20] attempt "to define a new field of study – the computational modeling of scientific behavior. Despite its relatively recent development, this research area has already made significant progress on issues that philosophers of science have traditionally ignored. In particular, the field has emphasized the nature of discovery rather than evaluation, and it has dealt with the *processes* that underlie science as well as the representation of knowledge. The result has been a rapidly growing set of computational models that deal with many facets of the scientific enterprise".

Computational science is a branch of cognitive science and of artificial intelligence that aims at constructing science by computational means. This defines the field of computational research on scientific discovery, which means "research on computational models of scientific discovery", i.e. the construction of "detailed computational models of the acquisition of knowledge in scientific domains" ([30, p.1-2]).

The implementation of AI methods for the discovery of scientific knowledge permits the automation of both inductive and abductive inferences. Over the past thirty years, machine learning scientists have provided us with computational systems that implement the rediscovery of many empirical laws of chemistry and physics, the automated discovery of equations in data bases, etc. Systems such as Bacon, PI, Dendral, Echo, IDS, Lagramge, Fahrenheit, Dalton etc. have been designed and applied for empirical discoveries. The cognitive scientists Herbert Simon and Paul Thagard and the computational scientists Pat Langley, Jeff Shrager, Bernd Hordhausen, Brian Falkenhainer, Jan Zytkow, D. Poole, A. C. Kakas, among others, have contributed to a general optimism about the possibility of automated inductive and abductive scientific discovery. For instance, under the theoretical assumption that chemical substances are made up of molecules, and molecules of atoms of the elements, and that atoms are conserved in reactions and that volumes of gasses, under constant temperature, are proportional to the numbers of their molecules, etc., dalton "deduces the chemical formulas of the molecules involved. For example, on being told that three volumes of hydrogen and one of nitrogen produce two of ammonia, it concludes, correctly, that hydrogen and nitrogen are H_2 and N_2, respectively, and that ammonia has the formula NH_3" [?] 10-11[]Sim1992.

Computational scientists repeatedly assert that they have filled the gap left behind by official philosophers of science, and they reproach them for abandoning the context of discovery ([31, p.23] and [32, p.3], [5, p.3, p.23, p.37-38], [30, p.20], [27, p.3]) Is the moment now ripe for the philosophers of science to regain the lost ground and to advance new possibilities in computational science? In order to answer this question, it is necessary to extend the spectrum of possibilities of scientific discovery. In this respect, both a wider concept of 'scientific discovery' and a closer attention to real scientific methods are unavoidable. In particular, we must investigate the possibility of new strategies for the introduction of novel ideas into science, beyond standard ampliative inferences such as induction and abduction.

As to the first issue, Dzeroski [27, p.3-4] define scientific discovery – briefly, and without further reference to exclusive methods of research – as "the process by which a scientist creates or finds some hitherto unknown knowledge, such as a class of objects, an empirical law, or an explanatory theory. ... A defining aspect of discovery is that the knowledge should be *new and previously unknown*" (my italics).

Let me announce the computational hypothesis of theoretical preduction. Computational preduction is tantamount to assuming the possibility of automating preduction mechanisms. The achievement of computational preduction is a task that must be carried out by computational scientists. The role of the philosopher of science is to signal an as yet unexplored possibility, namely that the research on computational models or systems of scientific discovery must be extended beyond ampliative inferences such as induction and abduction, and into the realm of anticipative inferences like theoretical preduction. This is the content of the computational hypothesis of theoretical preduction. If this hypothesis is worthy of consideration, then it is down to computational scientists whether or not they develop powerful deductive reasoners capable of putting it into practice.

References

[1] E. Böhm-Vitense. *Introduction to stellar astrophysics. Vol. 1: Basic stellar observations and data.* Cambridge: University Press, 1989.

[2] Bermúdez de Castro. *El Chico de la Gran Dolina. En los orígenes de lo humano.* Barcelona: Crítica, 2002.

[3] A. Eddington. *The Internal Constitution of the Stars.* Cambridge: University Press, 1926.

[4] A. Einstein. *Ideas and Opinions.* London: Souvenir Press, 1973.

[5] P. Langley et al. *Scientific Discovery. Computational Explorations of the Creative Processes.* Cambridge, MA: The MIT Press, 1987.

[6] N. R. Hanson. *Patterns of Discovery. An inquiry into the conceptual foundations of science.* Cambridge: University Press, 1958.

[7] G. Harman. The inference to the best explanation. *The Philosophical Review*, 74(1):88–95, 1965.

[8] G. Harman. Knowledge, inference and explanation. *American Philosophical Quarterly*, 5(3):164–173, 1968.

[9] J. R. Josephson and S. G. Josephson, editors. *Abductive Inference. Computation, Philosophy, Technology*. New York: Cambridge University Press, 1994.

[10] C. Lorenzo. Primeros homínidos. géneros y especies. In Eudald Carbonell, editor, *Homínidos: Las primeras ocupaciones de los continentes*. Barcelona: Ariel, 2005.

[11] L. Magnani. *Abduction, Reason and Science. Processes of Discovery and Explanation*. New York: Kluwer Academic/Plenum Publishers, 2001.

[12] L. Magnani and E. Belli. Abduction, fallacies and rationality in agent-based reasoning. In O. Pombo and A. Gerner, editors, *Abduction and the Process of Scientific Discovery*. Colecçao Documenta, Centro de Filosofia das Ciências da Universidade de Lisboa, Lisboa, 283-302, 2007.

[13] D. McKenzie. Plate tectonics: A surprising way to start a scientific career. In Naomi Oreskes, editor, *Plate Tectonic*. Boulder, Colorado: Westview Press, 2001.

[14] Peter Medawar. Hypothesis and imagination. In Paul A. Schilpp, editor, *The Philosophy of Karl Popper*, volume 14 of *Library of Living Philosophers*, chapter 7. Open Court Publishing Co., 1974.

[15] T. Nickles. Scientific discovery. In S. Psillos and M. Curd, editors, *The Routledge Companion to Philosophy of Science*. London: Routledge, 442-451, 2008.

[16] J. et al. Noonan. Sequencing and analysis of neanderthal genomic DNA. *Science*, 314, 2006.

[17] D. A. Ostlie and B. W. Carroll. *An Introduction to Modern Stellar Astrophysics*. Reading, MA: Addison-Wesley, 1996.

[18] C. S. Peirce. *Collected Papers*. Cambridge, MA: Harvard University Press, 1965.

[19] H. Putnam. *Reason, Truth and History*. Cambridge University Press, 1981.

[20] A. Ritter. Untersuchungen über die höhe der atmosphäre und die constitution gasförmiger weltkörper. *Annalen der Physik und Chemie*, 8:157–183, 1879.

[21] A. Rivadulla. Theoretical explanations in mathematical physics. In G. Boniolo et al., editor, *The Role of Mathematics in Physical Sciences*, pages 161–178. Dordrecht: Springer, 2005.

[22] A. Rivadulla. Discovery practices in natural sciences: From analogy to preduction. *Revista de Filosofía*, 33(1), 2008.

[23] A. Rivadulla. Ampliative and anticipative inferences in scientific discovery: Induction, abduction and preduction. In L. Fernández Moreno, editor, *Language, Nature and Science: New Perspectives*, pages 35–52. Madrid-México: Plaza y Valdés, 2009.

[24] A. Rivadulla. Anticipative preduction, sophisticated abduction and theoretical explanations in the methodology of physics. In J. L. González Recio, editor, *Philosophical Essays on Physics and Biology*, pages 351–364. Hildesheim-N.Y.: Georg Olms Verlag, 2009.

[25] A. Rivadulla. Complementary strategies in scientific discovery: Abduction and preduction. In *Ideas in Action: Proceedings of the Applying Peirce Conference*, volume 1 of

Nordic Studies in Pragmatism, pages 264–276. Helsinki: Nordic Pragmatism Network, 2010.

[26] A. Rivadulla. Abduction in observational and in theoretical sciences. Some examples of ibe in palaeontology and in cosmology. *Revista de Filosofía*, 4(2):143–152, 2015.

[27] P. Langley S. Dzeroski and L. Todorovski. Computational discovery of scientific knowledge. In Sazo Dzeroski and Ljupco Todorovski, editors, *Computational Discovery of Scientific Knowledge*, number 4660 in LNAI. Berlin-Heidelberg: Springer, 2007.

[28] John G. Sclater. *Heat Flow Under the Oceans*. Boulder, Colorado: Westview Press, 2001.

[29] H. Shapley. On the nature and cause of cepheid variation. *Astrophysical Journal*, 448, 1914.

[30] J. Shrager and P. Langley. Computational approaches to scientific discovery. In Jeff Shrager and Pat Langley, editors, *Computational Models of Scientific Discovery and Theory Decision*. San Mateo, California: Morgan Kaufmann Publishers, 1990.

[31] H. A. Simon. Scientific discovery and the psychology of problem solving. In R. G. Colodny, editor, *Mind and Cosmos. Essays in Contemporary Science and Philosophy*. Pittsburgh: University Press, 1966.

[32] H. A. Simon. Scientific discovery as problem solving. *International Sudies in the Philosophy of Science*, 6(1):3–14, 1992.

[33] P. Thagard. The best explanation: Criteria for theory choice. *The Journal of Philosophy*, 75(2):76–92, 1978.

[34] A. L. Wegener. *The Origin of Continents and Oceans*. New York: Dover, 1966.

[35] E. Zahar. Logic of discovery or psychology of invention? *Brit. J. Phil. Sc.*, 34:243–261, 1983.

Received January 2015

Course of Action Recommendations and Their Place in Developmental Abduction

Donna E. West
State University of New York at Cortland
westsimon@twcny.rr.com

Abstract

This inquiry identifies affective, cognitive, and linguistic evidence that, in point of fact, imperatives are responsible for the truly originary hypotheses clearly driven by instinctual abductions, given that the source for these instincts is often an external one. This account demonstrates how such imperatives beckon generation of proposition-making – recommending a course of action for self or other. This proposal likewise uncovers how children's event judgments acquire an action-based individual and social force – recommending a course of action for another. This is illustrated by children's need to recognize social and conversational roles prior to suggesting that others take those roles in particular event profiles (cf. [46] for further elaboration). It is obvious that children's imperative-making demonstrates the power to frame recommendations for future modes of conduct. Such includes: internalizing "pure" Secondness and arriving at percepts, translating them to perceptual judgments, and finally suggesting an entirely novel remediating path.

1 Introduction

This account demonstrates how recommendations for new courses of action constitute quintessential abductions, and how such recommendations are grounded (early on in ontogeny) in practical habits to feel, think, and act in particular ways. It asserts that in ontogeny a two-tiered process characterizes abductive rationality: receipt of unbidden guesses, and a stage in which their viability is determined illustrated by recommendations for would-be action habits. Accordingly, what drives inferential reasoning at the outset is an internal call to change an action habit (often

The author wishes to express gratitude to the anonymous referees for their energy and insightful comments.

motivated by instinct from external sources) rather than internally driven interrogative strategies constructed on the fly. The present model proffers that abductive reasoning emerges consequent to an impulse to resolve social and scientific problems via mental habits which explain inter- and intra- event relations. In turn, mental habits (regularities and their changes) inform abducers whether premises qualify as candidates to explain puzzling consequences. It is critical to note that the sign most instrumental in ascribing mental habits to relevant action correlates are indexes, since they naturally draw attentional paths between participant roles and the event structure in which they are likely to be imbedded. These attentional paths highlight regularities of states and action; and inferences are born with implicit suggestions of novel relations. Eventually, children's event repackaging constitutes implicit abductions, since new modes of logical syntax are applied to action performance. Children can implicitly draw inferences once iconic overlays are applied to indexes, bringing about the "habitation of events." At this stage in hypothesis-making, children can determine the suitability of particular inferential relations – whether to approve/dismiss the premise or whether to revise it. Linguistic directives, (in line with Peirce's and Vygotsky's models) propose explicit course of action recommendations, codifying and validating extended applications of courses of action in diverse context for different participants filling the same function. Tracing inferential reasoning from its inception as an instinct, through to its status as an organizing device is critical, if we are to recognize its significance in ontological pursuits. In short, primary to proposing novel courses of action is the impulse to express integral knowledge of "event morphology"—instinctual guessing regarding the participant-holder's orientations (epistemic, deontic).

Abductive endeavors require proposing successful solutions for potential participants in light of their past habits (preferences/conduct).[1] These factors determine the recommendation or suggested strategy for action in the face of unanticipated happenings; and their issuance via language imperatives serves to facilitate reaching the goal.

This account offers an alternative approach to best explanation rationality – it posits that anticipatory social logic and an imperative to act (how discrete persons should behave within event profiles) has its source in implicitly learned competencies (not in-born capacities) to guess correctly [42], and is inscribed upon ontological, more objective eco-based knowledge (in line with Magnani's [13, 15] and in press, and Gibson's 1979 models [6]), for novel plausible proposals to be formulated and

---

[1] While both Abduction and Retroduction appear for the first time in "Lessons of the History of Science" (c. 1896), and seem to be used interchangeably, Peirce deliberately abandons abduction as a term in 1907, preferring retroduction in a footnote in MS 318 "Pragmatism," cf. Bergman and Paavola's [2, p.1907] textual gloss in the Commens Dictionary of Peirce.

articulated at any point in the inquiry process. At certain junctures in the inquiry – impelled by "forced choice" [3, p.154], the function of abductions as imperatives surfaces – suggestions are proffered to improve another's approach to a problem. As Woods [51, p.367] intimates, the suggestions are subjunctively informed, in that they are constructed upon an appreciation for the legitimacy of individual perspectives, despite their uniqueness or lack of similarity to those of the abducer. Ontologically, forced choice relies upon a natural unfolding of potential causes and effects associated with events, particularly the unexpected consequence, in light of the imposition of distinctive physical and psychological phenomena for different event participants/observers.

These emerging, encapsulated recommendations are imperatives, materializing in a flash, resulting in retroductions – consolidations of past event memories and anticipatory propensities of participants and their roles. Retroductions which integrate how event morphology and social factors inform each other illustrate more than rudimentary, originary abductions; rather, they offer a novel avenue to inform others how to legitimately solve problems. Recommendations constitute an all-at-once suggestion (imperative) for future conduct for another, to be entertained and perhaps discarded in favor of a more fitting paradigm – ordinarily bearing some ultimate kernel of truth. Consonant with Magnani's [15, p.353-357] revisionary model of abduction, modifications to proffered hypotheses are effectuated consequent to salient but often tacit factors particular to each eco-cognitive framework, and may not represent the best strategy or explanation. In fact, "a best explanation" may never exist. The very existence of best explanations appears to violate Peirce's notion of the Final Interpretant, because best explanations already offer the ultimate operation, vitiating any need for a Final Interpretant. Under the best explanation paradigm, each abduction already represents the Final Interpretant – any further inquiry is tossed to the wind. The best explanation approach to abduction actually blocks the way of inquiry and the way to pragmatic relevance, which Peirce adamantly cautions against.

This account illustrates how abductions (spontaneous, plausible projections – revisionary in nature) are not best explanations, but the creative kind to which Magnani [16, p.267-268] and [14, p.28] refer. Findings which demonstrate the implicit nature of children's constructions of event profiles and their management of systems of imperative and subjunctive rationality will be examined to trace the ontogeny of abductive reasoning.

2 Originary Hypotheses- Their Application to Peirce's Categories

Especially with respect to originary hypotheses, Peirce introduces the issue of insight — sudden propositions developed in a flash[2]. This sudden aha-based instinct appears to be more in line with idiosyncratic compulsivity than are instincts grounded in physiological capacities (which are universal in nature). The former are Originary hypotheses rooted in the convergence of Firstness and Secondness, the intersection of possibility and brute force. Since, for Peirce, abduction is "the spontaneous conjecture of instinctive reason" [25, V.6 S.475], it necessarily goes beyond a physiological impetus—incorporating idiosyncratic percepts and discoveries derived from unique events in Secondness. The guessing instinct, which might better be named the guessing imperative (by which abductions are created), has its root in affective Firstness and Secondness; any selective or dimissory imperatives are rooted in logic. [] Tschaepe seems to conflate deliberate with conscious, as an abduction can be both unconscious and deliberate. More conscious abductions appear not to be as originary, but "revisionary," in Magnani's sense of the term. Peirce's later model further supports this distinction: between originary hypotheses (those generated out of sudden, unconscious insight); and those emergent from more regenerative, constructive and conscious effort. "Abduction moves from uncontrolled [automatic and originary] to controlled [conscious] [42, p.122]. The former kind of guessing (unconscious, instinct) translates into a habit of such proportion that it is realized in conduct which is less able to be regulated – propositions derived from these abductions are so automatic that they are tantamount to instincts. The more dynamic kind of abduction emanates from dynamic, deliberate and revisionary propositions, often conscious and less constrained by automatic processes.

Originary guessing illustrates Peirce's assertion that we have a propensity to guess correctly: it is tantamount to Firstness-based proposals intrinsic to perceptual judgments, and is formulated in the course of knee-jerk imperative responses; the remaining kinds of perceptual judgments are conscious, and require greater workmanship on the part of the inventor, which may likewise be guided by the imperative to guess right. According to Peirce, this "ability of guessing right is neither blind nor infallible, but is an *instinctive* ability, similar to the animal instinct of flying or nest-building of ordinary birds" ([25, V.6 S.476]).

Originary hypotheses can likewise be prompted by Thirdness-based imperatives. The impulse of doubt is just one of these, though without the ordinary face of Thirdness. Doubt can often dispel conventional propositions (elements defining

---

[2]"The abductive suggestion comes to us like a flash. It is an act of insight..." [25, V.5 S.181]

Thirdness), as in suspicions that a particular theory/algorithm no longer leads to the presumed end nor captures elements which later have augmented the original concept. Even young children display a propensity for this Thirdness-based imperative to take hold, e.g., in joint attentional linguistic exchanges – a child of 1;3 uses a banana as a telephone [21] and a child of 2;2 refers to a pencil as a hammer [50, p.319]. In fact, in the naming process, children regularly overextend the use of both nouns [18, 20, 19, 36, 37] and verbs [41] to incorporate unconventional members of the event type; and still later, they increase the number of slots within events to create event types [40, p.151-152], e.g., from "go," to "put" to "give."[3] "Go" encodes two slots (one for an animate self-start, migratory participant, the other for a spatial destination point); "put" accounts for three slots: an animate manually dexterous participant capable of intentionality, an inhabitable location with a surface, and a patient (ordinarily an object) typically incapable of self-start skills. "Give" invites still additional slots: ordinarily two manually dexterous participants with intentionality, a destination point (most often the location of one of the participants, and a transferable (tangible/intangible) commodity. The latter verb type profiles the dynamic, transcendent and reciprocal process of imposing goods or an attribute upon another – consonant with Gibson's [4, p.31] and [5, p.164-165] process of reafferent flow: "The living animal is stimulated not only from sources in the environment but also by itself. . . .Action-produced stimulation is *obtained*, not *imposed*—that is, obtained *by* the individual not imposed *on* him" [5, p.31]. Both afferent and efferent processes represent the character of the reafferent and on-line (instantaneously updated) perceptual account of a potential abducer: "Instead of entering the nervous system through receptors it re-enters. The input is not merely afferent, in the terminology of the neurologist, but *re*-afferent—that is, contingent on efferent output" Gibson [5, p.31]. Tracing the perceptual account can uncover the beginnings and development of perceptual judgments, in that it monitors dynamic changes in fields of output and reception within a system: "It is intrinsic to the flow of activity, not extrinsic to it; dependent on it, not independent of it" Gibson [5, p.31]. See [46] [47] for further discussion of the semiosis of abductive reasoning and event profiles.

These extensions and overextensions constitute abductions impelled by Thirdness, except that the imperative in Thirdness (to expand the original, conventional category) is not, in fact, conventional as Thirdness ordinarily is. Rather, it extends and alters conventional means/uses – reaching abductive status by virtue of new offerings: application to different movement valencies, participant attribute modifications (adding/deleting characteristics such as animacy/self-start), or simply by diversifying slot possibilities. Similarly, Thirdness is obviated in change of belief

---

[3]For further discussion, cf. [48].

127

processes, which (although latent) constitute subjunctive advances, integral to the semiosis of generating novel event profiles. Augmenting subjunctivity in event profiles motivates increased comprehension (albeit tacit) of degrees of certainty as to the likelihood of events to transpire, the likelihood of certain someones to take an active/passive slot in events, etc. To illustrate the issue of event certainty, doubt constitutes whether the other contributory events have materialized, whether the flow of contemporaneous events is sufficient to prompt other contingent events, or simply whether outlying participants/objects can fit into the original event profile – in view of the number and kinds of slots attributed to it. Hence, doubt can be assuaged (within the mind of the abducer or via the abduction itself) by inclusion of what may appear, at first glance, to be anomalous with respect to previously established propositional assumptions. As such, these extensions act in compliance with the imperative – to amplify original profiles. This form of revision represents a departure from Woods' treatment of "ignorance preservation," since the emergence of doubt and the introduction of new experience matrices, together with instantaneous reflection on them initiated by recommendation-based imperatives, result in significant affirmative changes to propositional logic. Such affirmative changes entail reorganization of mental schemes to accommodate novel, but not hasty assertions.

The influence of imperatives unquestionably energizes the semiosis of abductive reasoning – from their beginning in Firstness as prescinded elements of Secondness, to propositional reorganization in Thirdness. Accordingly, Peirce makes plane that abductions are frequently instigated by factors beyond Firstness and Secondness, despite his early misuse of "induction" (testing hypotheses) to define the process. In 1898, Peirce represents the operation of abduction as follows: "...[Induction] to be valid must be prompted by a definite doubt or at least an interrogation; and what is such an interrogation but first, a sense that we do not know something; second a desire to know it; and third an effort – implying a willingness to labor for the sake of seeing how the truth may really be" ([29, p.171]. Peirce's characterization of abduction here certainly reaches beyond posing questions and ignorance preservation to directives to produce novel templates. Since production of novel templates results (according to Peirce) in course of action recommendations, imperative based inducements are far from inconsequential, because their results are often tangible and can illustrate underlying inferential strategies. In fact, enacting a course of action recommendation can supply the impetus to organize behavioral steps within a framework to validate the strength of the recommendation in actual event scenarios.

As evidenced above, knowledge seeking is intrinsic to imperative directives which refine event structure. This is especially credible in view of Peirce's claim ([28, p.447]) that the propensity for "guessing right" (triggered by the imperative to investigate and codify kinds of events) is a universal capacity, even in some nonprimate species. Nonetheless, what is particularly noteworthy in the latter passage is his accentuation of the effect of Secondness upon the operation of abductions in Thirdness. His use of "interrogative" and "effort and ... willingness to labor for the truth," unequivocally demonstrate the relevance of Secondness to abductive processes in Thirdness; and the effort expended, coupled with a "willingness" supports Peirce's constructivist account. It is through the expenditure of energy via effort in Secondness (imposed by the imperative to notice particular components of experience in Secondness), together with the affect (in Firstness) which triggers the willingness to create and recreate novel hypotheses.

Peirce makes emphatic the originary nature of abductive reasoning – its idiosyncratic beginnings, out of the substance of assertions absent planning and deliberation. Despite the revisionary character of abductions (modified or dismissed) consequent to additional knowledge from anomalous C events, the instantaneous emergence of originary hypotheses should not be underestimated. In line with Tschaepe's [42, p.129] position, abductions would never materialize were potential abducers to engage in inquiry (interrogative processes) alone, without exacting some impulsive, truly insightful hunch flowing from the inner core of the individual abducer. In a word, the originary nature of abductions distinguishes them from interrogative phases; and the fact that Peirce brands them as originary unquestionably demonstrates both the influence of Firstness in the bargain, as well as the element of Secondness in constructing/fashioning explanatorily adequate guesses – the persistent labor which may be required to guard against hasty assertions. In fact, it is the imperative phase of abductions which precludes abducers from making hasty assertions when they are permitted to intervene to determine that the set of premises offered as a potential abduction has sufficient validity – that they are not hasty. The imperative serves as an index of the degree of accuracy present in the originary strategy. It communicates when the hypothesis is sufficiently well-formed to offer a viable course of action for the involved players. The imperative then has a regulatory function – indicating to the abducer whether the hypothesis is ready to be articulated. This regulative device limits defective judgments, whose premises were issued outside of our good guessing instinct or absent abeyance to the natural acumen of the guessing instinct (the imperative module). In short, with this imperative device (hastening guessing by determining when best to guess), fewer revisions are necessary, which lends economy, systematicity, and reliability to originary hunches in Thirdness.

It is as part of the system of originary guessing that Firstness "bridges the logical with the psychological," as Tschaepe [42, p.129] intimates. Firstness likewise encroaches upon Thirdness, bridging the phenomenological with the ontological, and the affective with the cognitive; new renditions are primarily fueled by Firstness, rejuvenating propositions and retroductions in Thirdness. Additionally, to reiterate: the influence of Secondness upon the semiosis of abduction is unparalleled – the impulsivity of imperatives, together with the expenditure of labor in repackaging retrospections into retroductions represents a quintessential illustration of its relevance. Peirce's concept of guessing (as originary hypothesis-making) is not founded upon flimsy assertions, independent of a search for the Final Interpretant; rather concerted effort to offer a viable, innovative protocol elevates productivity and work product. The purpose which drives the effort consists in elements of Firstness and Secondness – such that percussiveness gives rise to personalized output overlaying components of the stream of ever-flowing context. Here the confluence of Firstness and Secondness generates and regenerates "works of art" which often serve non-subjective ends. As such, guessing entails a deliberate process of choice and chance, typically without lengthy planning. This supports Tschaepe's [42] claim: "guessing is the initial deliberate originary activity of creating, selecting, and dismissing potential solutions..." [42, p.117]. Paavola [23, p.152] aptly describes the guessing instinct (development of originary hypotheses) as unconscious instances of problem solving. It appears obvious that although some intent to pursue the truth via our own competencies is operational and some awareness of how retrospections become retroductions, consciousness is ordinarily not inherent to this initial process. Accordingly, because of the lack of conscious intervention, abducers are often unable to reproduce/rearticulate the action-based recommendation; and even their effects (Interpretants) may evade notice. This evidences the fact that knowledge upon which abductions rest is often tacit in nature as Magnani [13, p.8] suggests. As such, originary abductions are frequently deliberate but unconscious primarily because they represent abductions in their initial stages. What Peirce explicitly attributes to an originary abduction is a primary or elementary novel set of premises holding together logically. It constitutes: "the simpler Hypothesis in the sense of the more facile and natural, the one that instinct suggests, that must be preferred" [25, V.6 S.477].

3 Further Application to Peirce's categories

Peirce characterizes reasoning from a pure Secondness perspective as "compulsive" ([25, V.2 S.268]). In fact, Peirce explicitly makes compulsivity a necessary component of action based and scientific discovery: "But how is it that all this truth has ever been lit up by a process in which there is no compulsiveness nor tendency toward compulsiveness" ([25, V.5 S.172])?

He articulates this same element, compulsivity to be a primary characteristic of all kinds of abduction, those effecting individual and collective ends. This compulsivity takes flight from the appearance on the scene of unexpected consequences. Part of the unexpected consequence is likewise spontaneous reactions which unforeseen circumstances impose. For Nubiola [22], abducers must resolve the doubt intrinsic to surprising events, such that they "regularize a surprising phenomenon to make the surprise disappear through the creation of a new habit" ([22, p.124]. Hence, Peirce's component of surprise, i.e., the surprising event, becomes a foundational imperative factor through Peirce's habit in the abductive turn. Orienting action within events for different players (inherent in recommending a course of action) demonstrates an obvious and sustained influence of Secondness upon hypothesis formation ([25, V.8 S.41]. "...[volition] does involve the sense of action and reaction, resistance, externality, otherness, pair-edness." Pure Secondness does not merely result in reaction to stimuli, but surfaces as "volition", having their foundation in Firstness – self initiated action. "Volition" here illustrates the import of imperative operations in abductive reasoning, in that volition provides the impetus to determine which responses from interrogative endeavors will be adopted and incorporated into retroductions and course of action recommendations.

To formulate a recommendation suitable to the particular complexion of the other in the specific context, children must ultimately supersede Secondness, and create an internal, virtual reality. They must assume event roles other than those which they have experienced (acting out the recommendation) and must apprehend the diverse nature of each event profile, e.g., agent, instrument, receiver, and the like by determining how others in distinct roles act out the recommendation (cf. [46, p.127], [45, p.164-165]). Nonetheless, Secondness is accorded a pivotal role in the operation of abductive reasoning, particularly when it is infused with an imperative – the volitional force driven by Firstness. Peirce accords direct experience a pivotal role in the emergence of higher reasoning skills ([25, V.8 S.266]: "The practical exigencies of life render Secondness the most prominent of the three. This is not a conception, nor is it a peculiar quality. It is an experience." The exigencies pregnant in experiences (both internal volitional and external ones intrinsic to the experience itself) provide the raw material upon which percepts and perceptual judgments are grounded. Hence, Secondness constitutes the rudimentary foundation for later decisions of what to attend to and how to direct others' attention, pivotal to recommending courses of action.

4   The Percept and the Perceptual Judgement

According to Peirce, "percepts come with beliefs, preconceptions, and prejudices leading to perceptual judgments; thus there is no hard and fast line of demarcation between perception, conception, interpretation and knowledge" ([25, V.5 S.184]. Likewise, while perceptual judgments are subject to criticism, percepts are not Short [35, p.512]. This is so since the latter is beyond an agent's control, agents of abductions can control the former. Although percepts may be far less subject to human control/intervention, they are not impervious to such agency. Because percepts are grounded in first impressions of sense [35, p.511] and are grounded in Secondness, they cannot qualify as abductions. Even when percepts rise to the level of interpretations – noticing similarities/differences with already experienced events, they fall short of abductive status, since although they may qualify as propositions, they lack any explanatory adequacy inherent in abductions ([35, p.517]. Elements of Secondness actively impinge upon perceptual judgements, suggesting the effort imposed by the agent in the abduction. Percepts can emerge consequent to deliberate attentional effort, contrary to Short's [35, p.518] claim, that perceptual judgments are made involuntarily, not requiring justification, insofar as they are predicated on a "look" ([25, V.7 S.627]). Since, according to Peirce, judgments are instinctual, they are not subject to justification since they are based on *a priori* knowledge ([25, V.1 S.118]). In fact, Short [35, p.520] capitalizes on two kinds of propensities which particularly represent instinctual judgments: "original knowledge in two instincts – the instinct of *feeding*, which brought with it elementary knowledge of mechanical forces, space, etc., and the instinct of *breeding*, which brought with it elementary knowledge of psychical motives, of time, etc." ([25, V.1 S.118]). This appears to dovetail with Hintikka's claim that abductions do not provide probabilistic support. In fact, an object/feature can be noticed as a result of particular preferences/salience for individual agents, or percepts may be couched in unconscious/automatic surveillance of certain arrays in particular ways. The fact that percepts/perceptual judgements are often unconscious does not relegate them to involuntary status, so as to disqualify them from consideration as potential abductions. In fact, the simple attentional act of looking toward a particular stimulus need not be involuntary as Short asserts.

5   The Emergence of Course of Action Rationality

According to Peirce: "All mental operations are of an inferential character" [?, 1:37]. This includes percepts, as well as perceptual judgements. In 1877 ([30, V.1 p.112]), Peirce takes this a step further by identifying the process which converts percepts and perceptual judgements to hypothetical status – accounting for how one inference (set of logically connected propositions) is ratified over others, namely, the governance of habit over thought and conduct. Kilpinen (forthcoming) identifies habit to be the "guiding principle" of abduction. In "How to Make Our Ideas Clear" ([30, V.1 p.131], Peirce illustrates the further application of habit to action and articulates that the "whole function of thought is to produce habits of action." Peirce's objective was not to refer to thoughts and actions which have already materialized, but which are about to materialize, those for which spontaneous conjecture is incipient. In the same passage, Peirce traces the process of hypothesis-making – proposing a set of affairs likely to be effectual in a context which has not yet materialized, but for which some potentiality exists: "[T]he ideal of the habit depends on how it might lead us to act, not merely under such circumstances as are likely to arise, but under such as might possibly occur, no matter how improbable they may be." The inception of abduction as course of action recommendations can be traced to Peirce's insistence that retroductions are not complete unless they "recommend a course of action," ([24, 1909 637:12]). To reach status as retroductions then, thought habits must be transferred into action habits.

The merit of this approach lies in its establishment of on-line proposals for discrete strategies within causative events and their sequence – to advance to the unanticipated consequence. In fact, without enactment, anticipation of unexpected consequences may well be stymied. The approach satisfies Peirce's Pragmatic Maxims, so as to make necessary action-base sequences to recreate the C event. It capitalizes on: the purpose for the recommended course of action, the means (via sequenced steps) to invoke the original surprising consequence. At the same time, recommending a course of action clarifies the problem-solving direction for the abducer, while providing increased resolve for those implementing the strategy, to follow the suggested course.

To successfully propose courses of action which lead diverse others to reproduce the unexpected consequence, children must initially rely upon subjunctive skills, such that reactions of distinct participants are anticipated to certain event types. Creative inferencing depends supremely upon two realizations: that events have different profiles/slots for participants which define social roles, and that a consequence of the same event affects others in different ways. In human ontogeny, recommendations for courses of action are prompted by subjective considerations; afterward they rely upon more objective perceptual judgments – how a reasonable person in the respective culture would react, or how a known unreasonable person might respond. Here Peirce informs us as to the ultimate purpose for propositions– the explanation for their existence – to "recommend a course of action" ([24, 1909 637:12]) functional to set the stage and trigger the C event. The recommendation is just that – a plausible recommendation, not an ultimate suggestion for another's conduct. In the same year [25, V.6 S.324], Peirce clarifies that the action which flows from recommendations is of the existential kind, which includes states of being and transitively based actions (acting upon something in a practical sense).

Peirce's characterization of abductions as "recommending a course of action" makes plain their dependence on perceptual judgments. Perceptual judgments entail inferences which qualify as abductions, in view of their propositional and classificatory nature. As such, to generate well-conceived hunches, children must rely upon consolidations of embodied experiences because they unite single actions, such that a law or habit is formed and propositions experience logical connections. These habits are aggregates of habit or as Stjernfelt [38, p.118] terms them, "action habits," which rest upon perceptual judgments, not merely single experiences guided by a percept or a percipuum. These "action habits" consist of implicit abductions (either of the instinctual or inferential kind); they constitute perceptual judgments, since their regularized schemes and embodiment of different persons in distinctive roles illustrate commonalities across similar event types. The influence of each contributing factor to the end-state is weighed; and a novel, adequate strategy is proposed, excluding immaterial factors and embracing factors according to their degree of influence on the event.

Moreover, the relevance of spontaneity and insight in forming new habits is obviated in recommending a course of action. Here Peirce illustrates how thought habit translates into action habit, completing retroductions by recommending a course of action. Peirce recounts a particularly poignant family incident in which his brother, Herbert, in a "flash" of "insight" made a determination to instantaneously cover a woman's burning dress with a rug to smother the flames, saving her from peril ([24, c.1902 5.538; c.1906 5.487 n1]. Herbert needed a course of action (a habit), to settle upon the best means of salvaging the person and dress after little opportunity to consider the effect of each factor in isolation. Possible competing but less effective remedies include: pushing the enflamed person onto the carpet, dousing her with water, removing her dress, etc. It is obvious that some form of instinct/insight initially served as the catalyst for the action recommendation, and that the insight was driven by a thought habit and a suggestion for action change. Although the inference need not amount to the best explanation or the best course of action, it must represent a plausible remedy. The explanation needs to convincingly evince a successful result given the surprising event (the igniting of the dress). The inference (to quickly cover the ignited item with a rug) qualifies as a viable abduction, likewise because it is not an outgrowth of empirical support/ deliberation (albeit fallible). Despite its fallibility, the premise qualifies as an abduction, given the likely success of its intervention for virtually any agent: "It is an act of *insight*, although of extremely fallible insight" [24, 1903 5.181]. The upshot is that abductions are subject to alteration/reformulation, given the potential of fallibility – they constitute abductions despite their incompleteness/overextension provided that they rest upon sound/plausible logic.

Without superseding elements of haecceity, inexperienced abducers would be hard pressed to generate abductions – to transcend self-participatory memories, and to (by virtue of their own impulses) go beyond and perhaps ignore irrelevant factors materializing in the context. Many irrelevant factors (which may not appear to be inconsequential from the outset) distract and mislead children; they constitute events which simply happen to co-occur in the context with the unanticipated event (which strikes children's notice). These event distractors are but one illustration of how salient contextual factors can compromise inference-building prior to 7;0 (Piaget and Inhelder [32, p.364]. It is this habit of attention to and reliance upon contextual exigencies that interferes with the semiosis of abductive reasoning. In short, children's recommendations regarding courses of action (however implicit and unfounded) would provide clear evidence of hasty assumptions, and would illustrate when abduction revisions are underway at early stages in ontogeny.

Peirce's explicit directive that abductions must "recommend a course of action" [24, 1909 637:12], although late in his attempts to communicate his concept of abduction, nonetheless is in line with his increasing pragmatic emphasis. "It will be remarked that the result of both Practical and Scientific Retroduction is to recommend a course of action," [24, 1909 637:12]. This recommendation is not generated consequent to extensive deliberation; rather it arises spontaneously [25, V.5 S.181]. This spontaneous reasoning emerges in a "flash" to preclude any contrivance from infecting the recommendation: "The abductive suggestion comes to us like a flash. It is an act of insight... "

There exists a double-tiered propositional scheme determining the recommendation for another based upon the eco-cognitive contexts which propel certain responses. This is consonant with Magnani's [15] approach, and provides an ignorance preserving character for the originator of the abduction – each reactor and each context actuates distinctive modes of action which include integration of affective, social, and logical factors relevant to an innovative action-centered proposal (cf. Magnani, Arfini, and Bertolotti forthcoming). As Magnani, et al. claim about affordance theory, "[it is] a conceptual tool, not only for investigating ... human perception, but also... human manipulation and distribution of cognitive meanings in suitable environmental supports." These proposals are quintessential illustrations of how ignorance is preserved because how another will respond in the same context as another and how the same individual will react in a different context remains undetermined. In these cases, abductions are integrative hypotheses to an ignorance problem—how a given unresolved issue is to be remedied.

Likewise paramount in recommending a course of action is subjunctive appreciation – a convincing suggestion which is functional for a known other or for an unknown, objective other (a reasonable person standard). Lakoff and Johnson's [12, p.34-35] stage theory, especially their characterization that "bodily projections" are necessary to more advanced applications of lived experience becomes particularly relevant to proposing plausible courses of action. This is so given that to propose behavior sequences capable of being successful, the abducer initially needs to insert self into the place of the other – measuring the combinatorial effect of all of the other's attributes. Without projecting one's self into the place of another, an outgrowth of lived experience, courses of action for potential experiencers are unlikely to be effectual. Later in development, once perspective-taking skills are substantially advanced and the reciprocal event roles are in place (West, 2013, 2014), the self need not be substituted for another; simply projecting the other into the new situation via imaginative skills suffices to predict that individual's response and to propose a recommendation accordingly. In short, projecting one's self into the situation links individual epistemological and deontic issues to social ones, perpetuating interpsychological and intrapsychological advances. Recommending a useful course of action particular to another (a remedy which is likely to function for another), requires metacognitive skills– to presuppose what others know, viz. the idiosyncratic emotional and informational-base of another.

Nevertheless, such expectations of another's affective and/or cognitive reactions give rise to creative, affirmative forces – capable of keeping others from harm. Essentially, well-formed expectations, predicated upon believable and well-founded retroductions, stimulate individual cognitive/epistemic growth (EP 2:192), and perhaps scientific advances, as well. This growth demonstrates a primary advance in intrapsychological reasoning, in that it materializes in the sudden synthesis of heretofore unforeseen connections/relations, to offer a novel rendition of or a projective account of events driven by instantiations of Firstness and unconventional Thirdness, without being subject to taboo or cultural sanctions. In fact, affect (which is an artifact of experience) is so crucial to epistemic development that, absent its influence, the inception of novel propositions which inhabit Peirce's sense of abduction are unlikely to come to fruition.

6 Peirce's Constructivist Approach

For Peirce, the abductive suggestion "is the idea of putting together what we had never before dreamed of putting together" [25, V.5 S.181]. While Peirce is silent as to the procedure employed to "put together" what we "never dreamed of putting together, he implies that the process is constructive, while at the same time, instinctual – generated in a flash." It is constructive, in that it is not an invention out of nothing, but is predicated upon forced selection of discrete elements never previously related logically; yet, connections are conceived of utilizing in-born capacities (naturally predisposed to individual discovery techniques) which develop into creative competencies.

Rather than advancing a purely innatist model, Peirce posits a more constructivist account of the ontogeny of logical rationality – contending that while in-born capacities are propensities common to the species, competencies are the skills developed consequent to the degree of capacity innately given. In other words, hunches which are grounded in inherent capacities, although universal, nevertheless are refined consequent to constructive endeavors/competencies—they (capacities) are provided in different degrees and kinds to each individual; and absent the operation of "putting together" pieces of implicit knowledge, via effort (competencies) in Secondness, propositions could never acquire sufficient novelty to qualify as abductions. If all abductions were required to be sequences of moves, already fabricated prior to engaging in the conduct which leads to the C event (as Hintikka indicates), many abductive moments would be overlooked, since the possibility of accounting for the impulses/capacities (internal, external) which initiate them would be excluded. Instead, Peirce's model of abduction incorporates opportunities to modify the sequence of behavioral advances, as well as to reject the initial form of the hunch in favor of a more workable hypothesis. In this context, the imperative is at work – taking advantage of the feel of enactments to charge the abducer with additional "what ifs." In Peirce's constructivist account, abductive reasoning is, unquestionably, an active process of stops and starts toward eventually hitting the target – the C event. This process toward hitting the target is demonstrated by Labra-Sproehnle [11, p.10] active, on-line reformation and dismissal of viable guesses is the substance of generating abductions. He describes an experiment in which subjects (ages beyond 7;0) play the game "Battleship," where the objective is to be the first to sink any of four ships given a set number of tries. Gleaning location information after each attempt is crucial for success; in fact, generating a strategy of moves prior to play enactment (according to Hintikka's approach) might well contribute to inferior performance. Of supreme import is the order in which each move is orchestrated with respect to the results of previous attempts to strike the battleship: "...it is assumed that the nature of the problem-solving behavior is connected intimately with the dynamic configuration of the continuum of results of the active thinking processes performed to solve the game."

Actively putting together fibers to solve previously baffling consequences (as Labra-Sproehnle illustrates) is consonant with Peirce's later approach. Peirce gave significant place to interrogative pursuits around 1900; but, thereafter he emphasizes active guessing –imperative paradigms directing agents toward the eventuality of viable hypotheses. In 1903, Peirce comments that abduction "merely suggests something that *may be*" [25, V.5 S.171-172]; indicating that the suggestion is an impulse, not a floundering sequence of questions. Playing the battle ship game requires a decision to venture forth with a proposition – that a portion of the opponent's craft is, in fact, at the point on the board where the missile strikes. This decision to strike at particular locations represents not a binary or open-ended inquiry, but an impulse that "it must be there." Peirce's characterization of a pheme[4] as "having a compulsive effect on the interpreter... " ([25, V.4 S.538], illustrates Peirce's commitment to imperatives as a type of command over the interpreter and perhaps over the consequences of the event as well: "Such a sign [pheme] intends or has the air of intending to force some idea either an interrogation or some action [in a command, or some belief in an assertion], upon the interpreter of it, just as if it were the direct and unmodified effect of that which it represents... " (c[24, 1906 295:6]. It is evident that a hunch determining the location of the opponent's craft in the battle ship game can easily surface as a "command," especially obvious given its spatial nature. Determining the location of an object (as in the case of the battle ship), relies upon substantive inferences regarding defining attributes, since knowledge of inherent iconic characteristics, e.g., shape, is necessary to strike the ship. Hence, in scenarios in which perspectives and placement are targeted, indexical and iconic inferences are the basis upon which resultant courses of action are founded. Symbolic representations of battle-shipping are far less material, such as knowledge of classificatory features; rather, indexical functions are primary, especially those driven by iconic representations. In short, the imperative function becomes heightened in hypothesis-making when the role of index in the showing enterprise supersedes that of other sign types. In fact, Peirce's increased recognition of the force of index in proposition-making and in illustrating the Dynamical Interpretant appear to coincide with his characterization of imperatives as signs advancing novel propositions. Accordingly, the imperative function becomes increasingly critical when attention to new assertions needs to be highlighted, as when targeting particular locations.

It is obvious that retroductions (gathering and consolidating previously unrelated

---

[4]Peirce initially employs pheme nearly interchangeably with Dicisign in MS [24, c.1906 295:6]. The substitution was intended to emphasize index's role as proposition tracer in the operation of inferential reasoning. It is that this time that Peirce launches index into its ultimate role as coordinator of perspectives, events, and episodes. As such, index traces the trajectory of modal propositions – indicative, to subjective imperative, to subjunctive (advocating diverse points of view), and finally to an objective imperative (advocating viable courses of action for the scientific/social community at large).

memories of past events) supersede the operation of simple inquiry – they convert single pieces of implicit knowledge (individual memories with their semantically laden associations) into a never-before-created fabric – an impulsively assembled mosaic, never before dreamed of according to Peirce. The imperative again rears its head by exacting delivery of individual retrospections from extinction. Their contribution to abduction constructs novel schemes informed by binding past event memories into a purposive consolidation (cf. [1] for a more complete account of binding). This account of retroduction illustrates that Peirce's teleological approach (every process has a purpose) is pervasive, even into the fabric of abductive logic; it entails combining individual retrospections into retroductions – creating a mosaic of increasing relevance of each retrospection to the others. In short, the process of retroduction (abduction) creates a kind of natural affinity between binding event memories and the means to guess correctly (the guessing instinct, as Peirce puts it).

Some additional clarity regarding Peirce's concept of instinct is in order, since he directly associates it with abduction, and since he uses the term to qualify guesses (abductions) as compulsive potentialities. Peirce is explicit that abductive reasoning is predicated upon triggering the flow of plausible hunches; and it is guessing which serves as the active manifestation of an imperative – that a proposition is about to surface. Essentially, guessing serves as the linguistic prompt from Vygotsky's perspective to comply with the command that retrospections are ready to intelligibly be bound with other retrospections into a retroduction (an abduction). In short, the guessing instinct represents a behavioral index that the imperative indeed has made its mark, indicating the point when the process of inquiry has reached momentary sufficiency.

Like the present approach, Paavola [23, p.152] distinguishes between abductive instinct and abductive inference – the former a rapid decision emanating from direct experience and the latter an organized search for hypotheses – but he appears to ascribe less credibility to instinct, which he assumes to be unconscious in all of its manifestations. Additionally, Paavola contends that there is "no sharp line" between them. Although Paavola [23, p.150-152] recognizes that instinct plays a role in the abductive process, he indicates that not every abduction relies upon instinct; abductions may, instead, derive wholly from a more organized strategy of analytic abductive inference in which agents explicitly search out premises. It is unclear what Paavola intends by "explicitly," whether he refers to a planned search toward identifying relevant antecedents, or whether he means that the knowledge base upon which abductive inferences are formulated is explicit. Moreover, Paavola does not provide additional hints as to its application – he is silent about any contrast between explicit and implicit knowledge/learning – the latter characterizing instinctual abductions. He does posit that despite their independent application, both types of abduction can be employed together in problem-solving pursuits [23, p.135]; but, he fails to demonstrate just how the two function in tandem – integrating instincts which may consist of flashes of insight from sources outside the self, in which personal experience is virtually irrelevant. Accordingly, Paavola's working definition of abductive instinct appears to frustrate a primary tenet of Peirce's intent, his agapistic principle[5] which is alive and well in abductive genres and especially potent in early childhood when experience is meager. If, in fact, abductive instinct were to depend wholly upon previous experience, as Paavola contends, the insight which, (according to Peirce, NEM III: 206) flows from "our dear and adorable creator" may in fact be thwarted.

The upshot is that Paavola fails to recognize that instinctive abductions cannot merely be unconscious, but likewise may begin from a seed of originary insight planted by a source outside the self. This regenerative impact of instinctual guessing upon abductive reasoning has the benefit of furthering Peirce's agapistic principle and his concept of the continuum—to imprint shared features across members of the continuum. This transfer of image function of instinctual abduction provides the key to how children transition from mere inquiry to developing plausible inferences. The process is not purely internal, nor does it always involve any conscious knowledge, however abbreviated. Rather, in this Peircean agapistic sense abductive instinct can emanate from something other than self-generated hunches; it can operate afterward to guide the abducer to select inferences which are not hasty, but which further the process of guessing right.

This interpretation supports Peirce's derivative claim that foundational to abductive reasoning is affective and subjunctive knowledge, materializing in one of three constructions: revised retroductions (consonant with Piaget's concept of accommodation), establishment of new propositions through the process of uniting unassociated retrospections, or simply a consequence of Firstness-based inventions, dreams/creative hallucinations/imaginings (cf. West in press for a further analysis of dreams and hallucinations as instinctual abductions). In short, it appears obvious that Peirce did not confound competency with capacity; without question, his use of instinct refers to competencies – supremely developmental and constructivist in nature, that admit unconventional but not counterintuitive nor internally inconsistent visional realizations. Nevertheless, Peirce's intent was not to disregard the necessity for universal capacities, while emphasizing the role of the constructivistic process for abduction. But, the latter process is more characteristic of novel logical assem-

---

[5]Peirce's Agapistic Principle is derived from his Kantian based assumption that man is made in the image of God (cf. [34, p.464]. This agapistic perspective naturally applies to abductive instinct and the propensity to guess right, in that not merely the conception of causal relations, but originary, novel insights spring unbidden.

blages. It entails gathering retrospections and reforming them into a significant, new pathway/habit via the process of retroduction. Essentially, this novel inferencing is a manifestation of what Peirce refers to as "habituescence" or "the consciousness of taking a habit," [24, 1913 930]. In short, Peirce's concept of instinct nicely incorporates the immediacy (response to an imperative) of the conjecture with the creative – capitalizing on regenerativity, while recognizing that instinctual guesses are likewise grounded in natural, perhaps unconscious propensities. In short, as a competency, original abductions rise to immediate productions of retroductions, while the less originary brand of abductions warrant a more deliberative process—application of a more dynamic, sequential system of constructive strategy-making.

7 Abduction as Imperative

In keeping with Peirce's repeated claim that abductions derive from natural instinctual hunches, this imperative-centered approach characterizes their initial phase as having a distinctly instinctual function, while in their latter phase, they require more deliberate (and perhaps conscious) consideration. In the first phase, the imperative functions to start the process of inquiry consequent to an insight which derives from a surprising or unexpected consequence; whereas, the imperative function in the latter phase involves exercising the "tendency to guess right" through self-talk. The latter is akin to Vygotsky's concept of inner speech, such that by intersubjective thought articulations (questions/possibilities) abducers formulate and direct hypothesis consideration by "guessing the right answer" Hintikka [7, p.527]. Hintikka [7, p.530] explains that abducers need to open up the probability of a question whose answer is anticipated.[6] Like the present approach, Hintikka recognizes two phases in abductive reasoning: guessing, which is "the only source for new information in an argument" [7, p.522], and abductive inference; but he gives special status to the latter, since it is in the latter process only, that the answer to interrogatives guides the hypothesis in a strategic manner. Nonetheless, Hintikka cautions that "such a strategic justification does not provide a warrant for any one particular step in the process," because every step may not "aid the overall aim of the inquiry" [7, p.530]. Hintikka's distinction between guessing (or as Peirce calls it, the "guessing instinct") and abductive inference is quite insightful, especially in recognizing that each step of the inquiry may not discernibly contribute explanatorily to the substance of the inquiry; but procedurally, each step is necessary insofar as it may "[open] up the

---

[6]As Hintikka further illustrates, "From the strategic vantage point we can say that thus any seriously asked question involves a tacit conjecture or guess" (1998:527). This abduction, founded upon less educated guesses, characterizes Peirce's "guessing instinct."

possibility of a question whose answer does so" Hintikka [7, p.530]. The present approach perceives the mechanics of opening up new possibilities both in its instinctual and inferential phases as an imperative, rather than an interrogative function, since it is structuring inner dialogue on the part of the abducer toward the perceived answer which organizes the steps toward making plausible hypotheses, not the questions themselves:

> The act of observation is the deliberate yielding of ourselves to that *force majeure*, - an early surrender at discretion, due to our foreseeing that we must, what we do be borne down by that power, at last. Now the surrender which we make in Retroduction, is a surrender to the Insistence of an Idea. The hypothesis, as the Frenchman says, *c'est plus fort que moi*. It is irresistible; it is imperative. We must throw open our gates and admit it at any rate for the time being [25, V.5 S.581] [29, p.170].

In fact, the ideas which impose themselves as imperatives upon our consciousness contract the questions which are eventually formulated. Both of these operations Peirce refers to as experiments, but experimentation of a particular kind—internally surrendering to ideas which compel themselves. Some of these ideas may be based on actual experience, while others may not. As Peirce clearly indicates, knowing the likely answer through intrasubjective (as opposed to intersubjective) dialogue constitutes abductions compelled by an imperative.

Unlike the present model, abductions are characterized by Hintikka [7, p.523] and Hookway [9, p.110], as question-answer steps, ignoring how interrogatives can be driven, in their originary state, by an imperative function, especially within an inferential process. Hintikka ultimately characterizes abductions as a series of perpetual inquiries (questions), and indicates that the sequence of interrogative steps is critical to reach success and to arrive at conclusions from surprising events. Although the order of questions within the inquiry is critical to assert a plausible explanation for the unexpected event as Hintikka contends, what impels the nature of the questions and their sequence in the first place, appears to be under-recognized – as a "goal" [7, p.527]. Despite the fact that Hintikka rightly acknowledges the need for a goal to direct the inquiry toward abductive closure, he fails to set forth the procedural paradigm for how questioning results in tenable hypotheses. The interrogative process itself is but a part of the over-all tendency to seek knowledge, an imperative. Peirce's 1908 letter to Lady Welby outlines in the trichotomies which attend to interpretants that "imperative" represents the second member within which "interrogative" is subsumed ([30, 1908 2:481]). In the draft of the 1908 letter, he specifically ascribes the function of the imperative to the Dynamic Interpretant ([30, 1908 2:483]). Peirce's stance regarding the ultimate plan for opening up new ground via the semiosis of interpretants measures imperative functions as higher in the hierarchy than are interrogative functions. As such, imperative functions have the power to impel the abducer by providing attentional and logical direction (like an index) – to envision the goal, together with some tacit means to ascertain such goal. In short, imperatives are ultimately responsible for the form and effectiveness of interrogative proposals, and hence for the plausibility of abductions. These imperatives are likewise responsible for determining whether hunches based upon instincts or retroductions qualify as abductions – if their results recommend a practical or scientific course of action. In fact, Peirce suggests that supplying a viable course of action recommendation is vital to success for abductive reasoning [24, 12:637]. In short, it is not inquiry or its sequence that determines the adequacy of explanations in the abductive turn, but determining when question-asking has momentarily satisfied/sated the new line of thought – when a workable hypothesis has been evoked.

The abductive process proffered herein departs from Woods' [51, p.365] ignorance preservation paradigm, in that it features imperatives to serve as gate-keepers of plausible hunches. In this role, imperatives preclude unfinished/incomplete hypotheses, and hasty assumptions from materializing as serious antecedents. Although Woods [51, p.370] defines abductions as suggestions for new actions, his claim that abduction is, unquestionably, an ignorance preserving proposition, actually subverts emergence of course of action recommendations. This is so since little impetus to proffer courses of action which could result in successful reproduction of the puzzling consequence are hardly likely to surface in the face of hasty and truncated hypotheses. Accordingly, Woods departs from Peirce's concept of abduction by failing to recognize the import of Peirce's Dynamical Interpretant to compel agents toward plausible hunches/good guesses. Woods' characterization of abduction as necessarily an ignorance preserving operation is a consequence of two misclassifications: underdetermining its ampliative effect, and implicitly assuming that all abductions are retroductions. He asserts that abductions must entail reasoning backwards from an unexperienced and surprising consequence to antecedent events which have necessarily been experienced by the abductor [51, p.370]. Woods [51, p.368] further manifests his mischaracterization of abductions as retroductions when he subsumes them along with two other obvious ignorance preserving operations: Subduence (accessing answers from sources other than internally driven hypotheses), and surrender (conscious substitution of inferior answers for less exacting ones). Gabbay and Woods' explicit assumption that abductions can only partially advance toward determining the answer [51, p.370], short-circuits any real effectiveness; abductions can never uncover the whole key to the inquiry – nor can they preserve truth.

In recognizing two distinctive but integratable abductive processes (instinct and inferential abductions), the present imperative-based model admits to the affirmative promise intrinsic to abductive reasoning – its unique capacity to "open up new ground" [27, V.III p.206 1907]. In fact, it is the imperative function which affords the abductive reasoning agent directional guidelines to select from among the pool of antecedents and to associate such with the natural process/flow toward the unexpected consequence. This process is guided by directionalized propositions. It assumes that hunches do not reach truly abductive status until the second phase, when retrospections are adequately integrated with one another to assert retroductions, the most uncomplicated, but weakest and least reliable form of reasoning:" I consider retroduction...to be the most important kind of reasoning, notwithstanding its very unreliable nature, because it is the only kind of reasoning that opens up new ground." [27, V.III: 206 1907]. Absent the indexical affordance provided by the imperative function (within both phases), questioning ad infinitum often end in logical and emotive quagmires which otherwise never become resolved. Without the impulse supplied by imperative operations, to "guess right," which as Peirce deems, leads us to the truth [30, 1902 2:250],[7] revising prior hypotheses to arrive at a more viable form of abduction is likely to be thwarted, since any new hunch consequent to revisionary hypotheses serves merely to diversify the state of ignorance. Moreover, it is the imperative function which triggers the revision of abductions, not, as Woods [51, p.367], and Thagard [39, p.244].[8] indicate, the subjunctive function, because while subjunctive functions can amplify hunches by providing diverse logical perspectives, they cannot supply the necessary orientation to determine which approaches need to be discarded in favor of more promising ones.

Furthermore, Hintikka's characterization of what constitutes abductions (unlike the imperative model proposed herein) overlooks another primary component of Peirce's paradigm – recommending a course of action. The force within the abducer to inform others how their habits of conduct are to increase the likelihood of the C event, is short circuited if the effect of proposing adequate strategies/approaches is unrecognized. Although Hintikka's contention has merit – that abductions are not truth-preserving in that they cannot provide even probabilistic support for their output [7, p.505] – his approach lacks adequacy to account for semiosis in the abductive process. His model fails to address dynamic change (advances) within the

---

[7]"...unless a man had a tendency to guess right, unless his guesses are better than tossing up a copper, no truth that he does not already virtually possess could ever be disclosed to him, while if he has any decided tendency to guess right, as he may have, then no matter how often he guesses wrong he will get at the truth at last" [30, 1902 2:250]

[8]"In sum, abduction is multimodal in that [it] can operate on a full range of perceptual as well as verbal representations. It also involves emotional reactions, both as input to mark a target as worthy of explanation and as output to signal satisfaction with an inferred hypothesis" Thagard [39, p.244].

question-asking process (which the imperative function supplies)—especially prominent in ontogeny, critical to determine Peirce's Final Interpretant. Inquiry Approach neglects the necessary eventuality of working toward refinement of abductions, to craft an hypothesis which can reliably (after intervals of inquiry and periodic invitations to express novel propositions) elicit the C event. Developing a series of revisionary hypotheses would be truncated, so too would be the semiosis of hunches, were abducers not induced to frame the proposal at each stage in the inquiry.

In fact, Vygotsky [43] is adamant that articulating proposed methods to ascertain an outcome is a major determinant in problem-solving success. Hintikka's model falls short of articulating the method; it merely validates judgments via particular sets of strategic principles. Hintikka's point is legitimate – that piecemeal "move by move rules" are not sufficiently systematized to lead to viable proposals [7, p.513]; but it ignores the impetus for continued inquiry. Despite the adequacy of Hintikka's claim, it fails to recognize abductions which glean insight from their implementation – the process of revising hypotheses after framing versions of the original proposal. As such, even the most revisionary abductions are governed by manipulative schemas (Magnani [15, p.374] which entail "thinking through doing and not only in the pragmatic sense about doing." Physical manipulation of action schemes (although they continue to influence abductive reasoning at later developmental stages) graduate to more mental manipulative schemes in which linguistic directives command courses of action. Magnani supports this premise: "The various procedures for manipulating objects, instruments, and experiences will be in their turn reinterpreted in terms of procedures for manipulating concepts, models, propositions, and formalisms" [13, p.55].

This process materializes via expressive language or simply by means of what Vygotsky [43, p.16-17] refers to as "egocentric speech" and "inner speech." The latter is a more advanced form of articulated speech; it constitutes an internal form of proposition development which transcends the need to self-regulate via external, audible directives from others (cf. [44, p.4-6] for an extended discussion). As applied to abduction (whether implicit or explicit), self-regulation and that from others' is essential, in that recommending courses of action are predicated upon articulated strategies impelling the enactment of interventions in problem-solving scenarios. When abductions emerge, they are often implicit, e.g., when gaze trajectories unite antecedents and their consequences; as such, they represent unarticulated propositions. Soon thereafter, inferences become explicit with the onset of language; and still later in development (when conditional terms are productive), inferences take on a more assertion-like character. Children's inferences are facilitated by articulated affirmations, because syntactic and semantic choices require them to package assumptions into some argument structure. As such, children are compelled to choose a particular antecedent or combination of antecedents to ensure production of the puzzling consequence. By mentally articulating a course of action, children select the more plausible premises and are set on a path to formulate a connection with the relevant consequence. This process does not enter the realm of induction, because children are not experimenting, merely formulating hypotheses via hunches which (like indexes) compel their attention. In short, Vygotsky's approach is consonant with that of Peirce, in that enactment, together with facilitating linguistic modes of directing remedial action represent the proposed courses of action, the essence of Peirce's abductive rationality [30, 1909 637: 12]. Additionally, when Peirce determines what children need to reach the "Supreme Art," commanding the self is the ultimate skill:

> 7th, how the facilitation asserted in the 4th point, where it is caused by attention to feelings, where the attention is of the nature of an inward exertion of power, is perhaps even greater when a different kind of exertion is substituted for the attention to feeling, this is different kind of exertion being describable to a person who has experienced it as an act of giving a compulsive command to oneself. Some books call it 'self-hypnotization,' whatever that may signify. This [is] effective whether there be any 'disposition,' i.e., any imperfectly developed or otherwise imperfect habit, or not" ([30, 1911 674:14].

This "inward exertion of power" to give "a compulsive command to oneself" is a skill without which the result of practical and scientific retroductions could not come to fruition, because without self-control or self- direction, embodiments for courses of action would be short-circuited. Consequently, the primary effect of retroduction would be unrealized without this imperative skill. The power of spontaneous propositional pointers in the form of these attentional commands to consolidate properties of events into propositions mapped onto the behaviors which embody them constitutes a powerful tool to generate viable assertions which propose an effective course of action, while revising strategies along the way. In this way, abductions are not piece-meal or fabricated sets of strategies, but hypothesis up-dates to be orchestrated on-line, in the stream of commerce. Such action streams may consist of conscious or unconscious organized bundles of goal-driven behaviors which coalesce to bring about an unexpected event. Although both conscious and unconscious strategy-making are deliberate, only the former is accounted for in Hintikka's model. This is so since to qualify as pre-formulated strategies (steps developed prior to exercising imperative enactments and language directives), generation of the set of steps before proceeding to reproduce the C event is required. Conversely, abductions derived from unconscious strategy-making often begin implementation before their combinatorial effect materializes. This unconscious kind typically consists of Burton's [3, p.153] notion of "forced choice" – guessing independent of experience, and is more effective than is guessing in the face of pure chance [3, p.153]. This type of operation is characteristic of children's early problem-solving endeavors, in that inferential reasoning materializes even in the absence of anticipated consequences, and often even without a relevant antecedent, despite their dependence upon the presence of tangibles.

8 Conclusion

Were these interrogative imperatives to stop there, the way to inquiry would be blocked, because conclusions and revised explanations would be thwarted. As such, inquiry might continue *ad infinitum* – without any impulse/compulsion to encapsulate new/surprising events into perceptual judgments, and to fold them into recommendations for future conduct. Despite the need for question-posing (perpetual knowledge-seeking) in the operation of abductive reasoning, it is insufficient to qualify as abductions, because the questioner must be impelled to feel that an information threshold has been reached upon which plausible hypotheses and suggestions for behavioral interventions can be drawn. Hence, imperatives surface to facilitate transcendence from interrogative processes to the formulation and expression of novel states of affairs from which recommendations for courses of action can issue. It is obvious then that abductive reasoning requires both interrogative and imperative operations. The latter is obviated by the fact that establishing foundations for courses of action for self or others entails a purposive boost – a reason for providing explanatory rationale and for designing action schemes tto implement pragmatic outcomes.

Nubiola's [22, p.125] position implicitly supports the need for an imperative module to restore well-formed guessing – to provide the push (affective and cognitive) necessary to have abducers readily share creative propositions with potential event participants. According to Nubiola, surprise within a surprising event is just that element; it "forces us to seek an abduction which converts the surprising phenomenon into a reasonable one" [22, p.125]. This surprise, coupled with the feeling of readiness to repackage events into behavioral strategies, together comprise the imperative module.

An imperative, rather than an interrogative triggers initial abductions because questions can be posed *ad infinitum* – perhaps without generating, selecting or settling upon a suitable premise. This imperative basis for abductions is constructed upon Peirce's requirement that abductions be spontaneous, such that they qualify as "forced guesses/decisions" [30, c. 1907 p.687]. The element of "force" upon which Burton [3] relies serves a managing function – that of expressing propositions which otherwise would remain hidden, never expressed, and ultimately lost.

Accordingly, this inquiry has identified affective, cognitive, and linguistic evidence that, in point of fact, imperatives are responsible for the truly originary hypotheses clearly driven by instinctual abductions, given that the source for these instincts is often an external, divine, one. This account demonstrates how such imperatives beckon generation of novel propositions – recommending a course of action for self or other. This proposal has uncovered how children's event judgments acquire an action-based individual and social force – recommending a course of action for another. This is illustrated by children's need to recognize social and conversational roles prior to suggesting that others take those roles in particular event profiles (cf. West 2014 for further elaboration). It is obvious that children's imperative-making demonstrates the power to frame recommendations for future modes of conduct [46, p.165-172]. Such includes: internalizing "pure" secondness and arriving at percepts, translating them to perceptual judgments, and finally suggesting an entirely novel remediating path [45, 46].

References

[1] A.D. Baddeley. *Working memory, thought, and action*. Oxford: OUP, 2007.

[2] M. Bergman and S. Paavola. *The Commens Dictionary: Peirce's Terms in His Own Words*. http://www.commens.org, 2014.

[3] R. Burton. The problem of control in abduction. *Transactions of the Charles S. Peirce Society*, 36(1):149–156, 2000.

[4] J.J. Gibson. The uses of proprioception and the detection of propriospecific information. In E. Reed and R. Jones, editors, *Reasons for Realism: Selected Essays of James J. Gibson*, pages 164–170. Hillsdale, NJ: Lawrence Erlbaum Associates, 1964/1982.

[5] J.J. Gibson. *The Senses Considered as Perceptual Systems*. Boston: Houghton-Mifflin, 1966.

[6] J.J. Gibson. *The Ecological Approach to Visual Perception*. Hillsdale, NJ: Lawrence Erlbaum Associates, 1979.

[7] J. Hintikka. What is abduction? the fundamental problem of contemporary epistemology. *Transactions of the Charles S. Peirce Society*, 34(3):503–533, 1998.

[8] J. Hintikka. *Socratic Epistemology: Exploration of Knowledge-Seeking by Questioning*. Cambridge: Cambridge University Press, 2007.

[9] C. Hookway. Interrogatives and uncontrollable abductions. *Semiotica*, 153(1/4):101–116, 2005.

[10] E. Kilpinen. In what sense exactly is peirce's habit-concept revolutionary? In D. West and M. Anderson, editors, *Consensus on Peirce's Concept of Habit: Before and Beyond Consciousness*. Heidelberg: Springer-Verlag, forthcoming.

[11] Fabián Labra-Spröhnle. The mind of a visionary: The morphology of cognitive anticipation as a cardinal symptom. In Mihai Nadin, editor, *Anticipation: Learning from the Past: The Russian/Soviet Contributions to the Science of Anticipation*, pages 369–381. Springer International Publishing, Cham, 2015.

[12] G. Lakoff and M. Johnson. *Philosophy in the Flesh*. New York: Basic Books, 1999.

[13] L. Magnani. *Abduction, Reason, and Science*. Processes of Discovery and Explanation New York: Kluwer Academic, 2001.

[14] L. Magnani. An abductive theory of scientific reasoning. *Semiotica*, 153(1/4):261–286, 2005.

[15] L. Magnani. *Abductive Cognition: The Epistemological and Eco-Cognitive Dimensions of Hypothetical Reasoning*. Heidelberg: Springer-Verlag, 2009.

[16] L. Magnani. Naturalizing logic. *Journal of Applied Logic*, 13:13–36, 2015.

[17] L. Magnani, S. Arfini, and T. Bertolotti. Of habit and abduction: Preserving ignorance or attaining knowledge? In D. West and M. Anderson, editors, *Consensus on Peirce's Concept of Habit: Before and Beyond Consciousness*. Heidelberg: Springer-Verlag., Forthcoming.

[18] E. Markman. How children constrain the possible meanings of words. In U. Neisser, editor, *Concepts and conceptual development: Ecological and intellectual factors in categorization*. Cambridge, Cambridge University Press, 1987.

[19] E. Markman and J. Hutchinson. Children's sensitivity to constraints on word meaning: Taxonomic vs thematic relations. *Cognitive Psychology*, 16:1–27, 1984.

[20] E. Markman and G. Wachtel. Children's use of mutual exclusivity to constrain the meanings of words. *Cognitive Psychology*, 20:121–157, 1988.

[21] L. McCune. A normative study of representational play at the transition to language. *Developmental Psychology*, 31(2):198–206, 1995.

[22] J. Nubiola. Abduction or the logic of surprise. *Semiotica*, 153(1/4):117–130, 2005.

[23] S. Paavola. Peircean abduction: Instinct or inference? *Semiotica*, 153(1/4):131–154, 2005.

[24] C.S. Peirce. Unpublished manuscripts. Unpublished manuscripts are dated according to the *Annotated Catalogue of the Papers of Charles S. Peirce*, ed. Richard Robin (Amherst: University of Massachusetts Press, 1967), and cited according to the convention of the Peirce Edition Project, using the numeral "0" as a place holder., 1866-1913.

[25] C.S. Peirce. *The Collected Papers of Charles Sanders Peirce 1866 - 1913*. Cambridge, Massachusetts: Harvard University Press, 1931-1935. Charles Hartshorne and Paul Weiss eds. (Vols. I-VI), Arthur Burks ed. (Vols. VII-VIII).

[26] C.S. Peirce. *Contributions to The Nation*. Lubbock: Texas Technological University Press, 1975-1987. K.L. Ketner and J.E. Cook eds.

[27] C.S. Peirce. *The New Elements of Mathematics*, volume Volume I Arithmetic, Volume

II Algebra and Geometry, Volume III/1 and III/2 Mathematical Miscellanea, Volume IV Mathematical Philosophy. Mouton Publishers, The Hague, 1976. Edited by Carolyn Eisele.

[28] C.S. Peirce. *The Writings of Charles S. Peirce: Chronological Edition*, volume IV (1879 - 1884). Bloomington: University of Indiana Press, 1989. C. Kloesel, et al. eds.

[29] C.S. Peirce. *Reasoning and the Logic of Things: The Cambridge Lectures of 1898*. Cambridge, MA: Harvard University Press, 1992. K. L. Ketner and H. Putnam eds.

[30] C.S. Peirce. *The Essential Peirce, Vols. 1 and 2*. Bloomington I.N.: Indiana University Press, 1992-1998. Peirce edition Project Eds.

[31] C.S. Peirce. *The Writings of Charles S. Peirce: Chronological Edition*, volume VI (1886 - 1890). Bloomington: University of Indiana Press, 2000. Peirce Edition Project.

[32] J. Piaget and B. Inhelder. *The Child's Conception of Space*. New York: W.W. Norton, 1948/1967. Translated by F.J. Langdon and J.L. Lunzer.

[33] J. Queiroz and F. Merrell. Abduction: Between subjectivity and objectivity. *Semiotica*, 153(1/4):1–7, 2005.

[34] Timothy Shanahan. The first moment of scientific inquiry: C,s, peirce on the logic of abduction. *Transactions of the Charles S. Peirce Society*, 22(4):450–466, 1986.

[35] T. L. Short. Was peirce a weak foundationalist? *Transactions of the Charles S. Peirce Society*, 36(4):503–528, 2000.

[36] N. Soja, S. Carey, and E. Spelke. Ontological categories guide young children's inductions of word meaning: Object terms and substance terms. *Cognition*, 38:179–211, 1991.

[37] N. Soja, S. Carey, and E. Spelke. Perception, ontology, and word meaning. *Cognition*, 45:101–107, 1992.

[38] F. Stjernfelt. *Natural Propositions: The Actuality of Peirce's Doctrine of Dicisigns*. Boston: Docent Press, 2014.

[39] P. Thagard. Abductive inference: From philosophical analysis to neural mechanisms. In A. Feeney and E. Heit, editors, *Inductive Reasoning: Experimental, Developmental, and Computational Approaches*, pages 226–247. Cambridge: Cambridge University Press, 2007.

[40] M. Tomasello. *The Cultural Origins of Human Cognition*. Cambridge, MA: Harvard University Press, 1999.

[41] M. Tomasello and S. Brandt. Flexibility in the semantics and syntax of children's early verb use. *Monographs of the Society for Research in Child Development*, 74(2):113–126, 2009.

[42] M. Tschaepe. Guessing and abduction. *Transactions of the Charles S. Peirce Society*, 50(1):115–138, 2014.

[43] L.S. Vygotsky. *Thought and Language*. E. Hanfman and G. Vakar (Trans.). Cambridge, MA: MIT Press, 1934/1962.

[44] D. West. Person deictics and imagination: Their metaphoric use in representational play. *California Linguistic Notes*, 35(1):1–25, 2010.

[45] D. West. *Deictic Imaginings: Semiosis at Work and at Play*. Heidelberg: Springer-Verlag, 2013.

[46] D. West. Perspective switching as event affordance: The ontogeny of abductive reasoning. *Cognitive Semiotics*, 7(2):149–175, 2014.

[47] D. West. Embodied experience and the semiosis of abductive reasoning. *Southern Semiotic Review*, 5(1):53–59, 2015a.

[48] D. West. The primacy of index in naming paradigms part i. *Respectus Philologicus*, 27(32):23–32, 2015b.

[49] D. West. Toward the final interpretant in children's pretense scenarios. In J. Pelkey, editor, *Semiotics 2015*. Toronto: Legas Press, In Press.

[50] D.P. Wolf. Understanding others: A longitudinal case study of the concept of independent agency. In G. E. Forman, editor, *Action and thought: From Sensorimotor Schemes to Symbolic Operations, 297-327*. New York: Academic Press, 1984.

[51] J. Woods. *Errors of Reasoning: Naturalizing the Logic of Inference*. London: College Publication, 2013.

 Received January 2015

Abduction: from the Ignorance Problem to the Ignorance Virtue

Tommaso Bertolotti
University of Pavia, Italy
`bertolotti@unipv.it`

Selene Arfini
University of Chieti-Pescara, Italy
`selene.arfini@unich.it`

Lorenzo Magnani
University of Pavia, Italy
`lmagnani@unipv.it`

Abstract

The ignorance issue afflicting abduction is usually intended as to what extent it can be seen as an "ignorance-preserving" (or "ignorance-mitigating") inference, and what this exactly means. In this paper, we will approach the issue of abduction and ignorance by leveraging the interplay of logical, epistemological and cognitive views on the matter. In detail, after introducing the ambiguous relationship between ignorance and achievement, we will deal with the presence of the ignorance element in the traditional logical formalizations of abduction, which configures the ignorance problem. Then, we will reframe the situation in an epistemological and cognitive perspective, making room for the view of ignorance as a virtue enabling the emergence of successful creative abductions.

Keywords: Abduction, Ignorance, Epistemology, Agent-Based Logics

1 Introduction

Scientia Potestas Est: Francis Bacon's famous quote, dating back to 1597, assessed a fundamental belief concerning the relationship between knowledge and power (firstly understood as a kind of *enablement*). Since the dawn of modern minds, human beings have been gathering and sharing knowledge because of its empowering role. Better and more extended knowledge concerning everything from the migrations of herds to the components of a smartphone have afforded better predictions and subsequently a better performance over the world.

Not only knowledge about objective external realities, such as scientific ones, plays an empowering role, but also knowledge about human beings gathered through experience or gossip and, for instance, through the fictional examples provided by literature helps making our social world more predictable – thus favoring those who actually posses this knowledge [16]. In brief, the claim that knowledge empowers is quite straightforward and there are not many occasions in which a human agent would claim that it is better to know less than to know more – obviously leaving aside all of those occasions in which, for emotional reasons, "we wish we hadn't known."

Yet, even before Gigerenzer's cognitive exploration of heuristics [7] and the emphasis on the catchy "less is more" effect (describing situations in which less knowledge actually affords a better performance than more knowledge), literature displayed a surprising number of instances seemingly arguing against the fact that *Scientia Potestas Est*. Such sometimes ambiguous relationship between knowledge, ignorance and achievement makes a quick appearance in Plato's *Theaetetus*, when the greek philosopher describes Thales falling in a well as he would stargaze and speculate while walking, so that a servant (the epitome of ignorance) makes fun of him.

More pregnantly, the Biblical book of Ecclesiastes reads as follows: "And I set my heart to know wisdom and to know madness and folly. I perceived that this also is grasping for the wind. For in much wisdom *is* much grief, And he who increases knowledge increases sorrow" (1:17-18). Sure the Bible refers to an existential kind of sorrow, but is it really that hard to interpret, *à la Gigerenzer*, "sorrow" in the sense of a failed achievement?

Another example: in a way that goes much beyond the epistemological virtue of simplicity (most appreciated by mathematicians and physicists), poet John Keats at the end of his *Ode on a Grecian Urn* stresses that "Beauty is truth, truth beauty, –that is all / Ye know on earth, and all ye need to know." To recommend the identity between beauty and truth as the only thing one has to know demarcates a very small subset of the equation between knowledge and power spelled out by Bacon more than two centuries earlier. To mention a more recent example, George Orwell's *Nineteen Eighty-Four* managed to depict a fictional, dystopian and yet disturbingly coherent society in which one of the governing maxims claims that *Ignorance is strenght*.

The fact that knowledge, ignorance and the achievement of a target do not necessarily come in constant proportions has recently acquired a new interest among logicians and epistemologists as it connects with what Hintikka [10] classified as the "fundamental problem of contemporary epistemology," that is *abduction*. Abductive reasoning has a crucial importance inasmuch as it is one of the few *ampliative reasonings*, as it allows the inferential expansion of the agent's knowledge beyond what she already knows.[1]

To make a straightforward example, if I know that 'all Romans are mortal' and that 'Caesar is mortal', I could try to expand my knowledge by abducing that 'Caesar is Roman': this is not necessarily true, because Caesar might very well be a pampered Chihuahua dog in Beverly Hills. Since the conclusion is not included in the premises (indeed it expanded my knowledge), it is not warranted by them either: to obtain it, I produced some knowledge out of something that was not knowledge yet, and that could be defined as *ignorance*. What about the relationship between the previous, underlying ignorance and the newly produced knowledge?

The ignorance issue afflicting abduction, as framed by [25] and [6], is about to what extent abduction can be seen as an "ignorance-preserving" – or "ignorance-mitigating" – inference, and what this means exactly.[2] In this paper, we will approach the issue of abduction and ignorance by leveraging the interplay of logical, epistemological and cognitive views on the matter. In detail, the next section will deal with the presence of the ignorance element in the traditional logical formalizations of abduction, which configures the ignorance problem. The third section will reframe the situation in an epistemological and cognitive perspective, making room for the view of ignorance as a virtue enabling the emergence of successful creative abductions.

---

[1]Abduction can be said to expand the agent's knowledge when 1) the knowledge-enhancing effect is at play and so the fruit of abduction is not potential knowledge but just knowledge (think of the Galilean thought experiment concerning falling bodies), and 2) when the guessed hypothesis (so potentially endowed with knowledge content) is accepted because it is evaluated (for example empirically).

[2][18] explains that abduction represents a kind of reasoning that is constitutively provisional, and it is possible to withdraw previous abductive results (even if empirically confirmed, that is appropriately considered "best explanations") in presence of new information. From the logical point of view this means that abduction represents a kind of nonmonotonic reasoning, and in this perspective we can even say that abduction interprets the "spirit" of modern science, where truths are never stable and absolute. Peirce also emphasized the "marvelous self-correcting property of reason" in general [21, 5.579]. So to say, abduction incarnates the human perennial search of new truths and the human Socratic awareness of a basic ignorance which can only be attenuated/mitigated. In sum, in this perspective abduction always preserves ignorance because it reminds us we can reach truths that can always be withdrawn; ignorance removal is at the same time constitutively related to ignorance regaining.

2 Abduction formalized and the emergence of the ignorance problem

The following schema is the basic structure of abduction provided by C. S. Peirce [21, 5.189]:

1. The surprising fact C is observed.

2. But if A were true, C would be a matter of course.

3. Hence there is reason to suspect that A is true.

Needless to say, the problem we are exposing in this paper is nested in the second step: *if A were true.* How to produce *A* if we assume that it was not *already* within our knowledge? With the characteristic *intellectual élan*, Peirce would also contend that human beings' capacity to make plausible abductive hypotheses is ultimately based on instinct. Albeit convincing from a number of perspectives, this description is question begging if one is trying to formalize the process at stake. How can we account for this? Instinct seems too non-explanatory, and all reasoning must happen according to some (leading) principles, but instinctive reasoning does not work that way, otherwise it would not be instinctive.

Such a conundrum about this seemingly strange reference, proposed by Peirce, to instinct as the ultimate basis of abduction can be easily clarified considering the following passage, where Peirce philosophically speculates taking advantage of an early "evolutionary" perspective:

> How was it that man was ever led to entertain that true theory? You cannot say that it happened by chance, because the possible theories, if not strictly innumerable, at any rate exceed a trillion? – or the third power of a million; and therefore the chances are too overwhelmingly against the single true theory in the twenty or thirty thousand years during which man has been a thinking animal, ever having come into any man's head. Besides, you cannot seriously think that every little chicken, that is hatched, has to rummage through all possible theories until it lights upon the good idea of picking up something and eating it. On the contrary, you think the chicken has an innate idea of doing this; that is to say, that it can think of this, but has no faculty of thinking anything else. *The chicken you say pecks by instinct. But if you are going to think every poor chicken endowed with an innate tendency toward a positive truth, why should you think that to man alone this gift is denied?"*[21, 5.591, added emphasis].

Furthermore, as shown by [17, chapter 5], the Peircean reduction of abduction to instinct is connected to his reduction of perception to a form of abduction. Peirce explains to us that perceptions are abductions, and thus that they are hypothetical and withdrawable. Moreover, given the fact that judgments in perception are fallible but indubitable abductions, we are not in any psychological condition to conceive that they are false, as they are unconscious habits of inference. Unconscious cognition legitimately enters the abductive processes (and not only in the case of some aspects of perception, as we will see). The same happens in the case of emotions, which provide a quick – even if often highly unreliable – abductive appraisal/explanation of given data, which is usually anomalous or inconsistent. Peirce also contends that perception is the fruit of an abductive "semiotic" activity that is inferential in itself. The philosophical reason is simple: Peirce stated that all thinking is in signs, and signs can be icons, indices, or symbols. The concept of sign includes *feeling, image, conception, and other representation*: inference is in turn a form of sign activity, that is, the word inference is not exhausted by its logical aspects and refers to the effect of various sensorial activities. We objected above that instinct seems too non-explanatory and surely all reasoning happens according to some (leading) principles, but instinctive reasoning does not take place according to some such reason: the problem is that instinct (and perception) do not perform abductions in the same way reasoning does, because Peircean theory of cognition is not restricted to inferential activities but refers to the effect of various other sensorial/semiotic activities.[3]

In summary, Peirce's idea that human beings' capacity to make plausible abductive hypotheses is ultimately based on instinct, which is in itself abductive, is extremely remarkable and tell us a lot about his modern philosophical perspective on human (and animal) cognition, informed by a kind of wide "cognitivism" *avant la lettre*: he says "It is a primary hypothesis underlying all abduction that the human mind is *akin to the truth* in the sense that in a finite number of guesses it will light upon the correct hypothesis" [7.220, added italics]. Albeit convincing from a number of perspectives, this description is question begging if one tries to formalize the process at stake.

Consequently, the best way to understand, and appropriately frame, the problem-

---

[3]The multifarious character of cognition is also testified by Peirce's conviction that *iconicity hybridates logicality*: the sentential aspects of symbolic disciplines like logic or algebra coexist with model-based features – iconic. Sentential features like symbols and conventional rules are intertwined with the spatial configuration, like in the case of "compound conventional signs" (written natural languages are concerned by iconic aspects too). What is called sentential abduction is in reality far from being strongly separated from model-based aspects: iconicity is always present in human reasoning, even if often hidden and implicit.

atic relationship between abduction and ignorance is to briefly review how abduction can be formalized. Whereas abduction can be of many kinds – visual, multi-modal, model-based and so on [17] – it is easier to set off from the kind of abduction defined as sentential, that is the one dealing with meaning expressed by a symbolic language to which propositions are associated, and then move to other kinds of abduction and see how ignorance affects them.

2.1 Framing the ignorance problem

The simplest way to understand abduction is the so-called "syllogistic" model, which sees abduction as the fallacy known as "fallacy of affirming the consequent."

P1: If A then B
P2: B
C: Then A

This model, albeit extremely straightforward, is relatively meagre in giving off details about how abduction actually works. The GW-model, proposed by Gabbay and Woods, is far more eloquent.

From this perspective the general form of an abductive inference can be formally rendered as follows. Let α be a proposition with respect to which you have an ignorance problem. Putting T for the agent's epistemic target with respect to the proposition α at any given time, K for his knowledge-base at that time, K^* for an immediate accessible successor-base of K that lies within the agent's means to produce in a timely way,[4] R as the attainment relation for T, $\rightsquigarrow$ as the *subjunctive* conditional relation, H as the agent's hypothesis, $K(H)$ as the revision of K upon the addition of H, $C(H)$ denotes the conjecture of H and H^c its activation. The general structure of abduction can be illustrated as follows, by recurring to the GW-schema (Gabbay & Woods):

---

[4] K^* is an accessible successor of K to the degree that an agent has the know-how to construct it in a timely way; i.e., in ways that are of service in the attainment of targets linked to K. For example if I want to know how to spell 'accommodate', and have forgotten, then my target can't be hit on the basis of K, what I now know. But I might go to my study and consult the dictionary. This is K^*. It solves a problem originally linked to K.

1. $T!\alpha$ [setting of T as an epistemic target with respect to a proposition α]

2. $\neg(R(K,T)$ [fact]

3. $\neg(R(K^*,T)$ [fact]

4. $H \notin K$ [fact]

5. $H \notin K^*$ [fact]

6. $\neg R(H,T)$ [fact]

7. $\neg R(K(H),T)$ [fact]

8. If $H \rightsquigarrow R(K(H),T)$ [fact]

9. H meets further conditions $S_1,....S_n$ [fact]

10. Therefore, $C(H)$ [sub-conclusion, 1-9]

11. Therefore, H^c [conclusion, 1-10]

A few notes about the G-W schema: basically, line 9. indicates that H has no more plausible or relevant rival constituting a greater degree of subjunctive attainment. Characterizing the S_i is the most difficult problem for abductive cognition, given the fact that in general there are many possible candidate hypotheses. It involves for instance the *consistency* and *minimality* constraints, corresponding to lines 4 and 5 of the standard AKM schema of abduction. The classical schematic representation of abduction is expressed by what [6] call AKM-schema, which is contrasted to their own (GW-schema). For A they refer to Aliseda [1; 2], for K to Kowalski [12], Kuipers [13], and Kakas *et al.* [11], for M to Magnani [14] and Meheus [19].[5] Finally, $C(H)$ is read "It is justified (or reasonable) to conjecture that H" and H^c is its activation, as the basis for *planned* "actions".

It is easy to see that the distinctive epistemic feature of abduction is captured by the schema. It is a given that H is not in the agent's knowledge-set. Nor is it in its immediate successor. Since H is not in K, then the revision of K by H is not a knowledge-successor set to K. Even so, $H \rightsquigarrow (K(H),T)$. So we have an ignorance-preservation, as maintained since the beginning of our discussion (cf. [25, chapter ten]).

Indeed, what matters for the preservation of ignorance, in the GW-schema, is that T cannot be attained on the basis of K. Neither can it be attained on the basis of any successor K^* of K that the agent knows then and there how to construct. H is not in K: H is a hypothesis that when reconciled to K produces an updated $K(H)$. H is such that if it were true, then $K(H)$ would attain T. The problem is

---

[5]A detailed illustration of the AKM schema is given in [17, chapter two, subsection 2.1.3].

that H is *only hypothesized*, so that the truth is not assured. Accordingly Gabbay and Woods contend that $K(H)$ *presumptively* attains T. That is, having hypothesized that H, the agent just "presumes" that his target is now attained. Given the fact that presumptive attainment is not attainment, the agent's abduction must be considered as preserving the ignorance that already gave rise to her (or its, in the case for example of a machine) initial ignorance-problem. Accordingly, abduction does not have to be considered the "solution" of an ignorance problem, but rather a response to it, in which the agent reaches presumptive attainment rather than actual attainment. $C(H)$ expresses the conclusion that it follows from the facts of the schema that H is a worthy object of conjecture. It is important to note that in order to solve a problem it is not necessary that an agent actually conjectures a hypothesis, but it is necessary that she states that the hypothesis is *worthy of conjecture*.

It is remarkable that in the above schema

> [...] $R(K(H), T)$ is false and yet that $H \rightsquigarrow (K(H), T)$ is true. Let us examine a case. Suppose that your target T is to know whether α is true. Suppose that, given your present resources, you are unable to attain that target. In other words, neither your K nor your K^* enables you to meet your target. Let H be another proposition that you don't know. So $K(H)$ is not a knowledge-set for you. On the principle that you can't get to know whether α on the basis of what you don't know, $K(H)$ won't enable you to attain T either. This is a point of some subtlety. Pages ago, weren't we insisting that there are contexts – autoepistemic contexts – in which not knowing something is a way of getting to know something else? No, we said that not knowing something was a way of getting to presume something else. But just to be clear, let us point out that in the GW-schema α and H are not candidates for the autoepistemic inference of α from H or $K(H)$. So $R(K(H), T)$ is false. $H \rightsquigarrow (K(H), T)$ is different. It says, subjunctively, that if H were true, then the result of adding H to K would attain T. Clearly this can be true while, for the same H, K and T, $R(K(H), T)$ is false [25, chapter eight].

It should be observed that the GW schema aims at illustrating the inferential structure of abduction, considering the agent's attempt to attain an epistemic target, her background knowledge and its possible (timely accessible) updated versions. Albeit the schema seems to suggest a fully explication of any abductive reasoning, it remains an explication of theoretical abduction at the sentential level; other types of abductive reasoning hardly fit this plain description (as manipulative abduction, model-based one, etc). Indeed there are traits of abduction, even at sentential level, that fall outside of a logical level of investigation and that need an eco-cognitive investigation to be analyzed. In this sense, the ignorance preservation remains within the background knowledge of the agent, because the activation of the hypothesis H does not modify K or $K*$, but simply helps to find an answer to one specific ignorance problem.

Finally, considering H justified to conjecture is not equivalent to considering it justified to accept/activate it and eventually to send H to experimental trial. H^c denotes the *decision* to release H for further premissory work in the domain of enquiry in which the original ignorance-problem arose, that is the activation of H as a positive *cognitive* basis for action. Woods observes:

> There are lots of cases in which abduction stops at line 10, that is, with the conjecture of the hypothesis in question but not its activation. When this happens, the reasoning that generates the conjecture does not constitute a positive basis for new action, that is, for acting *on* that hypothesis. Call these abductions *partial* as opposed to full. Peirce has drawn our attention to an important subclass of partial abductions. These are cases in which the conjecture of H is followed by a decision to submit it to experimental test. Now, to be sure, doing this is an action. It is an action *involving* H but it is not a case of acting *on* it. In a full abduction, H is activated by being released for inferential work in the domain of enquiry within which the ignorance-problem arose in the first place. In the Peircean cases, what counts is that H is withheld from such work. Of course, if H goes on to test favourably, it may then be released for subsequent inferential engagement [24].

We have to remember that this process of hypothesis evaluation (and so of hypothesis activation) is not abductive, but inductive, as Peirce contended. Woods adds: "Now it is quite true that epistemologists of a certain risk-averse bent might be drawn to the admonition that partial abduction is as good as abduction ever gets and that complete abduction, inference-activation and all, is a mistake that leaves any action prompted by it without an adequate rational grounding. This is not an unserious objection, but I have no time to give it its due here. Suffice it to say that there are real-life contexts of reasoning in which such conservatism is given short shrift, in fact is ignored altogether. One of these contexts is the criminal trial at common law" [24].

To summarize, we can say that the core of the ignorance-preservation problem of abduction is that the abducted hypothesis H is at best *minimal* and *consistent* with the rest of the agent's knowledge base K (in the rest of the article we will show how even this requirement is often rather wishful). H, though, did not belong to K nor to $K*$ in the beginning, and at the end even a satisfactory solution of the initial problem does not say why H should be derived from K or $K*$. Perhaps, as shown by [18], the initial ignorance is somehow *mitigated* but not totally removed.

Whereas the GW-model nicely captures "sentential" abduction, it is relatively obscure as to how H is generated (since it does not derive from K, it is hard to formulate a general mechanism for its production) and so it partially delineates an ignorance problem that is common to abduction in general. The issue regarding the *generation of H* is strictly related, as we will better see in the next section, to the difference between *selective* and *creative* abductive inferences: in the first case, the agent simply picks a hypothesis among a range of pre-formed or pre-selected ones, while in the second the abductive process accounts for the whole production of the hypothesis, as there are no previous ones – or they are unsatisfactory. This clearly opens up a necessary distinction in the kind of ignorance that is preserved, or mitigated: it can be either the ignorance affecting the selection of the correct hypothesis (e.g. "I don't know what train he caught" or "The doctors don't know if he's got a severe bronchitis or a mild case of pneumonia"), or an ignorance of a superior epistemological level, which cannot be reduced to an *ignorance-that*, concerning wholly uncharted epistemic domains.[6] To obtain a better view of this issue, and what is the role assigned to ignorance, we can turn to examine it from an epistemological and cognitive perspective.

3 The ignorance virtue in the cognitive epistemology of abduction

In order to get there, we must understand how a pivotal distinction between *theoretical* and *manipulative* abduction extends the application of the concept beyond a sentential dimension, but still referring to the agent's *knowledge-content*, or lack thereof. According to the point of view of C.S. Peirce, all thinking is in signs, and signs can be icons, indices or symbols. Moreover, all inferences are a form of sign activity, where the word sign includes "feeling, image, conception, and other representation" ([21, 5.283]) and, in Kantian words, all synthetic forms of cognition.

In order to clarify this concept, we should mention that the epistemological dis-

---

[6]As we will delineate better in the following section, what we called "ignorance-that" refers to specific ignorances the agent is aware of, against a broader kind of ignorance the agent is not aware of. In spite of the similarities, it does not relate to the difference between knowing-*that* and knowing-*how*.

tinction between theoretical and manipulative abduction is based on the possibility of (ideally) separating two aspects in real cognitive processes, resorting to the differentiation between theoretical/cognitive ones, where only "inner-neural" aspects are at stake, and manipulative ones, in which the interplay between internal and external aspects is fundamental. Theoretical abduction illustrates much of what is important in creative abductive reasoning, in humans and in computational programs: the objective of selecting and creating a set of hypotheses (diagnoses, causes, hypotheses) that are able to provide good (preferred) explanations of data (observations).

A more specified distinction divides this category in the aforementioned "sentential" abduction – which is related to logic and to verbal/symbolic inferences – and "model-based abduction" – which refers to the exploitation of internalized models of diagrams, pictures, etc.[7] Manipulative abduction, instead, accounts for many cases of explanations, occurring in science and in everyday reasoning, displaying a kind of "discovering through doing" in which the exploitation of the environment is crucial. Through manipulative abduction, new and yet unexpressed information is codified by means of manipulations of some external objects (epistemic and, in general, cognitive mediators). Manipulative abduction captures a large part of scientific thinking where the role of "acting" is central, and where the features of this "scientific acting" are implicit and hard to isolate: actions can provide otherwise unavailable information that enables the agent to solve problems by starting *and performing* a suitable abductive process of generation or selection of hypotheses ([17]).

---

[7]The distinction between off-line and on-line thinking is analyzed in detail in [17, subsection 3.6.5, pp. 189–193]. Some authors have raised doubts about the on-line/off-line distinction on the grounds that no thinking agent is ever wholly on-line or wholly off-line. We think this distinction is at least useful from an epistemological perspective as a way of theoretically illustrating different cognitive levels in human and animal cognition. It must be kept in mind that the theoretical distinctions between types of abduction are meant to frame the main traits of each type, but they are not necessarily mutually exclusive and different analyses may highlight different kinds of abduction at play in a same process. Sentential abduction refers to the possibility of working on *sentences*, be them expressed in logical or verbal language, and it is hence more closely connected to traditional logical studies. Nevertheless, the iconic dimension is never that far out: whereas the semantic understanding and appraisal of a sentence concern sentential abduction, the visual or auditive recognition of the signs expressing it rather involves a model-based approach to the process. Not to mention the language-forming capabilities afforded by diagrams, dicisigns, and icons. Peirce himself robustly exploited diagrammatic aspects of reasoning in his own research on logic: his invention of existential graphs is very well-known.

3.1 Abduction and lower cognitive processes

Our claim is that within the vast topic of abductive inference it is possible to exploit connections between high-level and low-level inferential patterns to obtain a better understanding of the different kinds of abductive cognition and thus to gain a better grasp on whether role of ignorance can be acknowledge as more than something that is necessarily "preserved" or at best "mitigated".

It is very hard for us, as human beings, to exit our language-based, *propositional-ized* framework. Furthermore, as we engage in the attempt to convey some meaning to each other (as we are doing right now, writing a paper), we cannot abstain from relying on a symbolic language endowed with propositional meaning. The "problem" is that we must recur to this kind of language also to describe events that occur in non-propositional terms (for instance at physiological, or neuro-chemical, or perceptive level). Such awareness, albeit raised to a slightly different scope, was already clear to Peirce himself who, speaking about perception, stressed the fact that when we think about perception we immediately turn our perceptual judgments into propositions, but this way we are not reflecting on raw perception anymore:

> Looking out of my window this lovely spring morning I see an azalea in full bloom. No no! I do not see that; though that is the only way I can describe what I see. *That* is a proposition, a sentence, a fact; but what I perceive is not proposition, sentence, fact, but only an image, which I make intelligible in part by means of a statement of fact. This statement is abstract; but what I see is concrete.[8]

It is clearly not possible to transcend the propositional level in theoretical communication, it would be very hard to communicate this argument making use of hormones and unmediated electric impulses, nevertheless we must be very careful not to let our "perceptual view" taint every conception of abductive inference.

Many kinds of abduction, in fact, do happen below the sentential-propositional threshold, which could be considered as a relatively new acquisition. Animal abduction [15], to make a clear example, involves forms of abductive inference that are clearly pre-sentential (as pre-linguistic) and can be individuated even in the "cognitive" faculties exhibited by bacteria reacting to their environments [3], not to mention the toolmaking ability displayed by crows which could be identified as manipulative abduction [23].

---

[8]Cf. the article "The proper treatment of hypotheses: a preliminary chapter, toward an examination of Hume's argument against miracles, in its logic and in its history" [1901] (in [22, p. 692]).

In these cases, following [8], we can say that cognition "shades off" into other kinds of biological processes. It is not usually considered as genuine cognition, for example, when some bacteria adjust themselves to changing circumstances around them by using little internal magnets to distinguish north and south and thus move towards water or when they use external clues, through tactile exploration, to adjust their metabolic processes: in this sense, it can be suggested that basilar forms of abductive inference indeed preceded the development of proper forms of cognition.

The reason for the warning we were calling for a few paragraphs above, against the overeagerness to turn everything into propositional language, is clear in this case. Even if we state that non-human animals and other organism are able to make more or less complex forms of abductions, we must be careful about the meaning we give to words such as *explanation, knowledge*, and subsequently *ignorance*.

We can nevertheless suggest that such inferential processes operate on kinds of *representation* that are produced abductively, and are most probably unapparent to the organisms making use of them.[9]

Of course, this whole argument could be affected by the cumbersome issue of the existence and nature of representations in the animal mind: we believe that from our purpose a *deflationary* approach can be fruitfully assumed. We could consider as a representation of the outside world any modification in the inner system that more or less corresponds to a modification of conditions in the outer world, and that can serve as a base for future behavior. Since cognitive agents are endowed with some kind of communication system (be it nervous and/or chemical), we can define mental representations as "patterns of neural [or chemical, or even genetic] activation," coherently with Clark's connectionist view [17, p. 160]. As for the (pragmatic) content of such representations, we believe that Millikan's insight is the most useful.

Millikan suggests that internal representations of animals might mostly consists of PPR (*"Push-me pull-you" representations*), meaning they are both aimed at representing a state of affairs and at producing another, often suggesting a chance for behavior as received by the Gibsonian/affordance tradition [20]. The indicative content of a PPR mental representation about external agent will therefore never be of the kind *Oh, look at that organism PERIOD* but rather *Look at that organism:*

---

[9]Although the notion of representation can be philosophically considered as "emptied" to a certain extent, we decided to maintain it for two main reasons: the contingent one, is to adhere to Millikan's authoritative lexicon as far as animal cognition is concerned; the more essential one is that this reflection belongs to the *model-based reasoning* framework, by which a model is *used* in order to achieve a goal. Representations, as models, do represent a target, and even if the same process can be conveyed by concepts such as "structural coupling of inner and outer systems," we feel that the notion of representation better depicts its instrumental role.

should I attack/avoid/hurt/kill/eat it/mate with it?:

> An animal's action has to be initiated from the animal's own location.
> So in order to act, the animal has to take account of how the things to
> be acted on are related to itself, not just how they are related to one
> another. In the simplest cases, the relevant relation may consist merely
> in the affording situation's occurring in roughly the same location and at
> the same time as the animal's perception and consequent action. More
> typically, it will include a more specific relation to an affording object,
> such as a spatial relation, or a size relative to the animal's size, or a
> weight relative to the animal's weight or strength, and so forth [20, p.
> 19].

Millkan's contention is that animal representations are bases for action. This
comes as no striking news, because any cognizant is wired so to proceed from representation to action in order to survive. The presence of a central controller is not
needed to explain why some abductive representations are followed by actions and
some not: we can hypothesize that while perception stimulates the activation of a
neural network, only if the electrochemical signal reaches a certain threshold it can
"fire" the activation of a distinct motor-related neural network, triggering aggression
or escape. Such process can easily happen without the presence of a central intelligence that assesses the representation and decides when action should be enacted.
In this sense, many kinds of low-level abductions could be seen as *self-performing
abductions*, handling ignorance to their agents' unawareness.

This is not true of animals alone, but also many of the abductive inferences
human beings operate are of this kind, from perception (of inner and outer states
of things) upwards. Conscious, sentential abductive inferences are just the highest
steps of the pyramid. Indeed, to say that an organism can detect the presence of
another organism does not compel us into affirming that it has *consciousness* of the
other organism's presence: it suffices to imagine that the states of neural activation
originate a mental representation fit to guide its behavior.

With the same reasoning, we cannot always speak about "ignorance" as a matter
of *unattained sentential propositions about the world*: this is all the more true in
case of the ignorance preserved in abductive inferences performed by agents with
or without the capacity to explain their behavior at a sentential level. It will be
sufficient to sat that they lack the comprehension of how some facts affect their
action-reaction behavior. But it is worth mentioning that the ignorance-preserving
trait remains unaltered by the different formal expressions with which we can express
the inferences. The relationship between abduction and ignorance is "locked" in

both its formal expression and in its cognitively relevant occurrences; the difference between abductive processes (and their conditions of effectiveness) are connected to the different interpretations of ignorance the agent is preserving, and exploiting, during the abductive process.

In order to investigate the specific forms of ignorance mitigated (and preserved) while performing an abductive inference, we will follow the division between selective and creative abduction we mentioned before, inasmuch as it concerns the epistemological dynamics and the generation of hypotheses within or outside the knowledge of the agent. In the next section, we will provide an eco-cognitive exposition of hypothesis-generation presented through the two kinds of abduction and the forms of ignorance preserved and/or mitigated during the processes. This scheme will show the specific difference between the simple *preservation of ignorance* the selective abduction implies (which is the cause of its broad usefulness but also of its unquestionable handicap in the scientific practice), and the enhancement of knowledge *through ignorance* brought about by the generation of a new hypothesis (daring but functional) in a creative abduction.

3.2   Creative abduction and the virtuous prerogative of ignorance

Abduction, as an inferential activity, is obviously performed when the agent is embedded in a constant dynamic of action-reaction with her surroundings. This can also be seen as a negotiation of signs and data that she is catching and diffusing throughout the epistemic process. In the cognitive economy of the agent[10] abduction does not only concern a certain amount of known information, but also the endowment of some signs with a practical activation (as the PPR representations and affordable values). Thus, adbuctive processes causes the development of the agent's cognitive environment through the transformation of some unexploited data into something unexpectedly useful. These performances are strictly related to what the agent knows better about, and what is far from her natural, usual or established field of expertise.

Simplifying, we can see the complex of data the agent possesses and manages, together with those that are within her reach in her cognitive environment, as an agent-centered system. Her topics of expertise, the knowledge she usually employs, correspond to the *central information*: she can easily attain them, and she has a minimal ignorance about them. Instead, the information that is still within the agent's cognitive system but that is not in her dominion of expertise, or that she is vaguely ignorant about, correspond to the *peripheral information*: she knows

---

[10]"An economy is an ecology for the generation and distribution of wealth. A cognitive economy is an ecology for the generation and distribution of knowledge" [25, p. 85].

something about it but it is not part of her practical knowledge field. Abductive inferences are performed upon and within the cognitive economy of the agent, acting on these two parts of her system, albeit in two very different manners and making use of two very different kinds of ignorance.

The first type of ignorance is set within the limits of the agent's cognitive environment and it is rooted in her own central information. It involves a part of delusion about the actual knowledge the agent has on her field of expertise (which is kind of natural to expect). In order to overcome it, all it takes is a specific question, but also the agent's awareness about the information she lacks. When the agent knows what is missing, she can obtain the answer through some targeted questions; when her knowledge does not cover the amount of data she thinks it does, something is missing in her central information. **Performed in this context, abduction placates the ignorance that the agent recognizes: this prompts the inference in the first place, and maintains what she cannot expect there.** The richness of this kind of inferential activity depends on the agent's interest and competence.

By doing this, the agent enacts a *selective abductive inference*. It gives the agent the possibility to inquire into her specific ignorance and find the *best explanation* [17], selecting it from a counted number of choices (or still within a type of possible choices): the hypothesis still preserves the ignorance about the unforeseen possibility that it could be less than the best possible explanation. Think of a doctor struggling with her ignorance on whether the patient is affected by a severe bronchitis or a mild pneumonia: a new symptom can solve this ignorance, help the doctor formulate her diagnosis and start the treatment, but of course the diagnosis could still be wrong, and the doctor could shift from a small ignorance between two possibilities to a wrong course of action.[11]

---

[11] As contended by [18], if we say that truth can be reached through a "simple" abduction (both selective or creative) where simple means that it does not involving an evaluation phase, which coincides with the whole inference to the best explanation, fortified by an empirical evaluation, then it seems we confront a manifest incoherence. In this perspective it is contended that even a simple abduction can provide truth, even if it is epistemically "inert" from the empirical perspective. Why? We can solve the incoherence by observing that we should be compelled to consider abduction as ignorance-preserving only if we consider the empirical test *the only way* of conferring truth to a hypothetical knowledge content. This clause being accepted, in the framework of the formal model of abduction introduced above the ignorance preservation appears natural and unquestionable. However, if we admit that there are ways to accept a hypothetical knowledge content different from the empirical test, simple abduction is not necessarily constitutively ignorance-preserving: in the end we are dealing with a disagreement about the nature of *knowledge*, as Woods himself contends. Those who consider abduction as an inference to the best explanation – that is as a truth conferring achievement involving empirical evaluation – obviously cannot consider abductive inference as ignorance-preserving. Those who consider abduction as a mere activity of guessing are more inclined to accept its ignorance-preserving character. On this matter, also refer to footnote 2.

It is obviously possible to go beyond one's ignorance within the central information, but in order to do that the agent has to face what is missing and perform an abduction like the one just described. Obviously it is an enhancement in the management of her eco-cognitive structure, but it is again a more or less ignorance-preserving inference: as we saw, the agent "only" becomes aware of what she recognizes to be missing.

Conversely, the second type of ignorance is harder to manage than the first. It does not require just a specific question to be inquired, and so discovered. Indeed, it does necessitate more than the agent's ordinary expertise in order to become apparent: rather, it requires more patience and resources to be integrated within the core of her knowledge base. In order to discover a way to attain one's target inside this kind of ignorance, it becomes necessary to change the agent's eco-cognitive system and enhance it with the perspective that even in the most peripheral part of her knowledge base there are still plenty of useful answers to her questions. It also involves a change in the direction of the *interest* that it supposed to guide the abductive process.

Here it is not possible to use a selective abduction to direct the inquiry within such a vast and problematic ignorance. The method that can shed some light is Magnani's aforementioned *creative abduction* ([17]), and in particular the *trans-paradigmatic abduction* ([9]).

In these cases the hypotheses "transcend" the vocabulary of the evidence language, as opposed to the cases of simple inductive generalizations: the most interesting case of creative abduction is called by [9] *trans-paradigmatic* abduction. This is the case where the fundamental ontological principles provided by the background knowledge are violated, and the new hypothesis transcends the immediate empirical agreement between the two paradigms, for example in the well-known case of the abductive discovery of totally new physical concepts during the transition from classical mechanics to quantum mechanics [17].

Creative abduction – and its trans-paradigmatic form above all [9] – does not provide a simple selection of hypotheses but, through the change of the eco-cognitive paradigm, it provides a brand new field to investigate. When the agent cannot afford a specific question, or method of enquiry, because she cannot describe what she does not know – which is indeed unaffordable for her – it becomes necessary to perform a creative *context-shift* for example through an almost serendipitous creation of an alternative pattern.[12]

---

[12]By advocating serendipity we do not refer to a total randomness in the process of context shifting: conversely, we refer to its partially unpredictable and *emergent* nature. Even serendipity itself refers to the background, skills (and peculiar ignorance) of an agent – consider Louis Pasteur's famous saying that "In the fields of observation luck favors only the prepared mind."

Thus, enquiring within the second type of ignorance opens the possibility to discover a whole *hypothesis-cluster*, leading to the possibility of attaining the target in one or more new ways. Creative abduction, in this sense, exploits the ignorance of the agent rather than eliminating it. It is a way to claim knowledge and data by exerting a powerful epistemic drive through the ignorance of the agent.

The development on a new hypothesis-cluster signifies that, out of vast ignorance, one will not just draw a particular abductive answer (for instance whether San Diego is larger than San Antonio), but develop a new vocabulary, a new syntax and a new grammar. That was the case in the big example, already mentioned, of Einstein developing the Relativity Theory, but one can think of different examples too. Consider a case in history of technology: many readers will be acquainted with the (somewhat frustrating) history of handheld and PDA keyboards. Until the first first half for the years 2000s, the arm (or finger!) race was about developing the best typing accuracy, with on-screen keyboards with or without stylus, physical keyboards (folding or sliding) with hard or soft keys, optional keyboards in different sizes and so on. Yet, the typing accuracy deriving from fitting twenty-something keys in an area slightly bigger than a tiny notepad could not be overly improved. The new cluster-hypothesis, so far nested in the developers' ignorance, was to concentrate not on the typing but on the processing of that typing, developing predictive software that could overcome the mistakes caused by physical constraints and learn from the habits of the users. That meant moving from a typewriter mindset to an artificial-intelligence one. It was not one single hypothesis (for instance involving the best shape of a thumb-keyboard), but an hypothesis-cluster opening up a new field of research and development, unforeseen a few years earlier but from which there was clearly no going back.

This kind of solution to an ignorance issue provides an enhancement of the agent's perspective and knowledge but its consequences are wider than in the cases previously described, that is when one's precise ignorance is concerned (overcoming my ignorance about the capital of California, and learning that it is Sacramento, might improve my local performances but it is unlikely to constitute a system-shattering hypothesis-cluster). Obviously, the epistemic risk and opportunity at stake involved by a daring hypothesis, in the Popperian sense, are high. But the opening of a new field of research, of development or simply of reasoning is already a possibility toward compensating errors, misevaluations, or to further improve the most promising components.[13]

---

[13]Even if the issue is clearly related to the topic at stake, we are not especially dealing with the pragmatic relationship between ignorance and courses of action where ignorance can be a synonym of *uncertainty*, as in the case of economics. A specific investigation about that is due soon, setting off from some reflections introduced in this paper. In the meanwhile, some other exploratory ideas

4 Conclusion

Abduction, as the ignorance-preserving inference, has a complicated connection with knowledge and ignorance. The history of the concept, as well as its employment as a coherent tool for explaining lower and higher level of cognition, testifies to the richness of this entanglement and surely advocates for the benefits it produces. As we repeatedly said, and coherently with the Peircean heritage, abduction feeds upon signs but also upon the unexpected possibilities that those signs can provide to the cognitive prepared agent. Highlighting the epistemic *value* of ignorance not just as something that is preserved or mitigated is a crucial step before one can achieve, and then implement, the computational modeling of creative inferences based on clusters of hypotheses produced from within an agent's ignorance – not considering the latter as a simple lack of knowledge anymore.

References

[1] A. Aliseda. *Seeking Explanations: Abduction in Logic, Philosophy of Science and Artificial Intelligence*. PhD thesis, Amsterdam: Institute for Logic, Language and Computation, 1997.

[2] A. Aliseda. *Abductive Reasoning. Logical Investigations into Discovery and Explanation*. Springer, Berlin, 2006.

[3] E. Ben Jacob, Y. Shapira, and A. I. Tauber. Seeking the foundation of cognition in bacteria. From Schrödinger's negative entropy to latent information. *Physica A*, 359:495–524, 2006.

[4] T. Bertolotti. *Patterns of Rationality: Recurring Inferences in Science, Social Cognition and Religious Thinking*. Springer, Berling / Heidelberg, 2015.

[5] T. Bertolotti and L. Magnani. Contemporary finance as a critical cognitive niche: An epistemological outlook on the uncertain effects of contrasting uncertainty. *Minds & Society*, 2015. Forthcoming.

[6] D. M. Gabbay and J. Woods. *The Reach of Abduction*. North-Holland, Amsterdam, 2005. Volume 2 of *A Practical Logic of Cognitive Systems*.

[7] G. Gigerenzer and H. Brighton. Homo heuristicus: Why biased minds make better inferences. *Topics in Cognitive Science*, 1:107–143, 2009.

[8] P. Godfrey-Smith. Environmental complexity and the evolution of cognition. In R. Sternberg and K. Kaufman, editors, *The Evolution of Intelligence*, pages 233–249. Lawrence Erlbaum Associates, Mawhah, NJ, 2002.

[9] F. V. Hendricks and J. Faye. Abducting explanation. In L. Magnani, N. J. Nersessian, and P. Thagard, editors, *Model-Based Reasoning in Scientific Discovery*, pages 271–294. Kluwer Academic/Plenum Publishers, New York, 1999.

can be found in [4, chapter 9] and the forthcoming [5].

[10] J. Hintikka. What is abduction? The fundamental problem of contemporary epistemology. *Transactions of the Charles S. Peirce Society*, 34:503–533, 1998.

[11] A Kakas, R. A. Kowalski, and F. Toni. Abductive logic programming. *Journal of Logic and Computation*, 2(6):719–770, 1993.

[12] R. A. Kowalski. *Logic for Problem Solving*. Elsevier, New York, 1979.

[13] T. A. F. Kuipers. Abduction aiming at empirical progress of even truth approximation leading to a challenge for computational modelling. *Foundations of Science*, 4:307–323, 1999.

[14] L. Magnani. *Abduction, Reason, and Science. Processes of Discovery and Explanation*. Kluwer Academic/Plenum Publishers, New York, 2001.

[15] L. Magnani. Animal abduction. From mindless organisms to artifactual mediators. In L. Magnani and P. Li, editors, *Model-Based Reasoning in Science, Technology, and Medicine*, pages 3–37. Springer, Berlin, 2007.

[16] L. Magnani. *Morality in a Technological World. Knowledge as Duty*. Cambridge University Press, Cambridge, 2007.

[17] L. Magnani. *Abductive Cognition. The Epistemological and Eco-Cognitive Dimensions of Hypothetical Reasoning*. Springer, Heidelberg/Berlin, 2009.

[18] L. Magnani. Is abduction ignorance-preserving? Conventions, models, and fictions in science. *Logic Journal of the IGPL*, 21(6):882–914, 2013.

[19] J. Meheus, L. Verhoeven, M. Van Dyck, and D. Provijn. Ampliative adaptive logics and the foundation of logic-based approaches to abduction. In L. Magnani, N. J. Nersessian, and C. Pizzi, editors, *Logical and Computational Aspects of Model-Based Reasoning*, pages 39–71. Kluwer Academic Publishers, Dordrecht, 2002.

[20] R.G. Millikan. On reading signs: Some differences between us and the others. In D. Kimbrough Oller and U. Griebel, editors, *Evolution of Communication Systems: A Comparative Approach*, pages 15–29. The MIT Press, Cambridge; MA, 2004.

[21] C. S. Peirce. *Collected Papers of Charles Sanders Peirce*. Harvard University Press, Cambridge, MA, 1931-1958. vols. 1-6, Hartshorne, C. and Weiss, P., eds.; vols. 7-8, Burks, A. W., ed.

[22] C. S. Peirce. *The Charles S. Peirce Papers: Manuscript Collection in the Houghton Library*. The University of Massachusetts Press, Worcester, MA, 1966. Annotated Catalogue of the Papers of Charles S. Peirce. Numbered according to Richard S. Robin. Available in the Peirce Microfilm edition. Pagination: CSP = Peirce / ISP = Institute for Studies in Pragmaticism.

[23] A. A. S. Weir and A. Kacelnik. A New Caledonian crow (corvus moneduloides) creatively re-designs tools by bending or unbending aluminium strips. *Animal Cognition*, 9:317–334, 2006.

[24] J. Woods. Ignorance, inference and proof: abductive logic meets the criminal law. In G. Tuzet and D. Canale, editors, *The Rules of Inference: Inferentialism in Law and Philosophy*, pages 151–185, Heidelberg/Berlin, 2009. Egea.

[25] J. Woods. *Errors of Reasoning. Naturalizing the Logic of Inference*. College Publica

tions, London, 2013.

Received April 2015

www.ingramcontent.com/pod-product-compliance
Lightning Source LLC
Chambersburg PA
CBHW051411200326
41520CB00023B/7191

9 781848 902145